The Urban Book Series

The Urban Book Series is a resource for urban studies and geography research worldwide. It provides a unique and innovative resource for the latest developments in the field, nurturing a comprehensive and encompassing publication venue for urban studies, urban geography, planning and regional development.

The series publishes peer-reviewed volumes related to urbanization, sustainability, urban environments, sustainable urbanism, governance, globalization, urban and sustainable development, spatial and area studies, urban management, transport systems, urban infrastructure, urban dynamics, green cities and urban landscapes. It also invites research which documents urbanization processes and urban dynamics on a national, regional and local level, welcoming case studies, as well as comparative and applied research.

The series will appeal to urbanists, geographers, planners, engineers, architects, policy makers, and to all of those interested in a wide-ranging overview of contemporary urban studies and innovations in the field. It accepts monographs, edited volumes and textbooks.

Now Indexed by Scopus!

More information about this series at http://www.springer.com/series/14773

Vítor Oliveira
Editor

Morphological Research in Planning, Urban Design and Architecture

Springer

Editor
Vítor Oliveira
Engenharia Civil, Faculdade de Engenharia
Universidade do Porto
Porto, Portugal

ISSN 2365-757X ISSN 2365-7588 (electronic)
The Urban Book Series
ISBN 978-3-030-66459-6 ISBN 978-3-030-66460-2 (eBook)
https://doi.org/10.1007/978-3-030-66460-2

This Springer imprint is published by the registered company Springer Nature Switzerland AG
The registered company address is: Gewerbestrasse 11, 6330 Cham, Switzerland

Foreword

The exploration of the interface between the academic study of urban morphology and the practical application of some of its findings has gained considerable impetus in recent years. The present volume marks a significant step in the progress of this endeavour in two ways. First, from a philosophical perspective, several of the contributions validate the fundamental significance of urban morphology to interventions in the urban landscape. This significance is easy enough to assert but a reasoned theoretical basis is necessary to provide a platform for any meaningful operative intervention. Several chapters in this book provide such a foundation. Second, each of the chapters demonstrates instances of tangible interventions where urban morphological principles, refined and developed through scholarly research, have been applied to concrete aspects of urban design or townscape intervention.

This book represents the culmination of years of debate, even soul-searching, within urban morphological circles. The emergence of the 'research-practice' interface as a major consideration for urban morphology is difficult to date with any precision but its antecedents may be found in some of the early ISUF Conferences. For example, in the final plenary session at Lausanne in 1996 it was suggested that the next Conference (Birmingham 1997) could consider 'professional themes, focusing on common problems faced by practitioners' (Kropf 1997) and effort should be made to involve developers, politicians and others in related fields. But within a few years Whitehand (2000) could still reflect that 'it is evident that some of the disciplinary and other divisions are insufficiently permeable' and note that in relation to links between urban morphology and architectural and planning practice, 'The two activities remain in largely separate realms'.

Nevertheless, examples of the application of morphological ideas to urban design and practice are not just a recent phenomenon. A rapid search through the pages of some early issues of 'Urban Morphology' reveals that a number of instances can be found. Just two decades ago, Tony Hall (2000) lamented the prevalence of two-dimensional representation (and therefore, thinking) in development proposals and the absence of consideration of three-dimensional aspects. This 'serious flaw' inhibited the promotion of mixed-use, sustainable urban forms. Hall proposed a morphologically based matrix of levels of intervention to guide outcomes in redevelopment areas. Within the same issue of 'Urban Morphology', a planning practitioner—Louise Thomas (2000)—argued that morphological analysis was a primary design tool and presented a six-point illustration of how the examination of morphological diagrams could aid future planning and design decision-making. However, her main conclusion was that 'the techniques of urban morphology still need a great deal of development in order to be widely useful in urban design' (Thomas 2000, p. 34). Beyond a comment that urban morphology needed to focus less on history and more on the forces shaping the future, there was no further indication of what form such developments could take. Around the same time, McGlynn and Samuels (2000) developed an operational method of how to introduce the morphological approach to residential builders. Possibly one of the clearest illustrations of the application of morphological research to a practical, planning situation was demonstrated by Marco Maretto in his study of urban renewal in the historic centre of Chioggia in the Venetian Lagoon area (Maretto 2005). This study showed how the 'tools' of urban morphology and building typology could be used as the basis for a sensitive urban redesign project in an area of extreme historic and architectural value.

These examples suggest that ideas and, indeed, practical instances of the application of urban morphological principles to actual cases of urban design and development have not been lacking in the past. In fact, in 2001 Malfroy commented that 'In reality, opportunities for collaboration between specialists in urban morphology and architects or planners are not so rare as the small number of published accounts might lead one to believe' (Malfroy 2001, p. 63). A comment that, perhaps, reveals much about the operational norms of the development process within different national contexts.

This point had already been highlighted by Samuels (1990) in the context of architectural practice, noting that urban morphology and building typology studies were an essential part of the training and technical skills of Italian architects but virtually absent in their British colleagues. Thus, whilst we can cite early examples of attempts to bridge the academic urban morphology—practical urban development gap, these are fairly isolated cases. Furthermore, in the past, there has been no lack of practical suggestions as to how urban morphology could aid the design process, but cases of their application remained limited. It was dissatisfaction with this situation that led to the creation of the ISUF Task Force on Research and Practice in January 2012, chaired by Ivor Samuels. A particular concern of this group was with the issue of 'how' knowledge generated by researchers in urban morphology could actually reach practitioners (Samuels 2013). A number of recommendations of this Task Force still lie on the table, but debate remained vigorous as evidenced *inter alia* by a number of contributions to the 'Viewpoints' section of 'Urban Morphology', Volume 17 (1). If not exactly leading the agenda, the issue of the research and practice divide has never been far from the surface in the minds of urban morphologists. What has been lacking is a comprehensive, accessible statement of the 'state of the art' and this book provides exactly that. It represents the culmination of two decades-worth of efforts to bridge the research and practice gap.

The question of why urban morphologists have devoted so much time and energy to the question of the relationship between research and practice is an extremely interesting one and is capable of being answered in different ways and at different levels. If not always explicit, many of these are dealt with implicitly in this book. But we may ask—what are the key characteristics of academic urban morphology that convince its adherents that they have a role to play in the practical application of their research findings? The features mentioned below are not necessarily exclusive to urban morphology and they are generic in character but they provide the basic infrastructure upon which the discipline rests. First, the subject matter of urban morphology is 'grounded', it is concerned with tangible elements in the urban landscape, elements which are visible and blend together to form a visual experience and it is that visual experience to which the community relates in various ways. Second, the 'founding fathers' of morphological study have bequeathed 'frameworks for analysis'—they have provided a system or systems and methodologies for the detailed study of urban form, surely a necessary preliminary to any intervention? This does not mean to say that these frameworks should be slavishly followed. The point is that they provide a starting point, or possibly a structure for critical analysis of the problem under consideration. A key component of such frameworks is the notion of a 'hierarchy' of morphological elements or components at the intra-urban and inter-urban scales, proceeding, for example, from individual plots through morphological regions to plan types on the much larger scale of a cultural region. It is the combination of these elements within a hierarchical structure that constitutes much of the distinctiveness of urban forms but is so rarely acknowledged by practitioners. Less frequently recognised and often latent, rather than studied directly, is urban morphology's inherent concern with the 'aesthetics' of the townscape. The individual experience of the townscape has a 'spontaneous emotional dimension' (Conzen 1981, p. 82). In a rarely cited treatise, MRG Conzen drew attention to 'the emotive function of the environment needed by society'. Although fundamental, particularly in terms of well-being, this factor is all too often trumped by much shorter-term commercial considerations. A further, and growing, characteristic of urban morphology is the concern with 'cross-cultural studies' and/or 'comparisons'. At the simplest level, it surely must be the case that the practice of urban design and the creation and change of urban form in any one locale has much to learn from elsewhere. Despite the remarkable contradiction of the growth of ease of communication being paralleled by the growth of insular, 'fortress' mentalities, with its strong roots in geography, urban morphological research has much to offer practice in exploring different solutions to shared urban problems. Finally, urban morphological research has an enviable reputation for 'rigour, precision and detail'. In fact, this has arguably been one of the obstacles to the creation of closer relationships between research and practice. But, in an age of 'spin' and of largely ideologically driven 'symbolic' gestures and policy 'initiatives', the rigour and precision that characterise both scientific exploration and effective intervention, is surely the main weapon in countering the conveniently superficial and the downright false.

This book constitutes the most significant development so far in this long-running attempt to demonstrate the relevance of urban morphology to the applied fields of architecture, planning and urban design. It provides significant evidence of progress and it demonstrates, in one place, the wide range of environments and elements of the built form where urban morphological research has been applied. What it also demonstrates is that it is not enough for urban morphologists to simply assert their relevance and claim that they 'could' play a part, if only they were asked. Echoing Louise Thomas's comment earlier in this Foreword, urban morphologists still need to do more themselves to demonstrate their relevance. There has perhaps been, in the past, rather too much of a sense of the 'accusatory' in the urban morphologist's claim that architects, planners and urban designers are ignoring them. Urban morphologists need to 'show' their relevance—and that is precisely what this book does.

Newcastle upon Tyne, UK Michael Barke

References

Conzen MRG (1981) Geography and townscape conservation. In: Whitehand, JWR (ed) The urban landscape: historical development and management. Academic Press, London, pp 75–86

Hall T (2000) How morphology can improve development plans. Urban Morphol 4:29–32

Kropf K (1997) ISUF 1996: Third international seminar on urban form, Lausanne, Switzerland, 24–27 July 1996. Urban Morphol 1:66

Malfroy S (2001) Urban morphology and project consulting: a Berlin experience. Urban Morphol 5:63–80

Maretto M (2005) Urban morphology as a basis for urban design: the project for the Isola dei Cantieri in Chiogga. Urban Morphol 9:29–44

McGlynn S, Samuels I (2000) The funnel, the sieve and the template: towards an operational urban morphology. Urban Morphol 4:79–89

Samuels I (1990) Architectural practice and urban morphology. In: Slater TR (ed) The built form of western cities. Leicester University Press, Leicester, 415–435

Samuels I (2013) ISUF Task force on research and practice in urban morphology: an interim report. Urban Morphol 17:40–43

Thomas L (2000) Urban morphology in practice. Urban Morphol 4:32–34

Whitehand JWR (2000) Editorial comment: from explanation to prescription. Urban Morphol 4:1–2

Contents

Abbreviations

ABF	*Architectes des Bâtiments de France*
BMPC	Beijing Municipal Planning Commission
CABE	Commission for Architecture and the Built Environment
CAUE	*Conseil d'Architecture, d'Urbanisme et de l' Environnement*
COS	*Coefficient d' Occupation des Sols*
DCLG	Department for Communities and Local Government
DDE	*Direction Départemental de l' Equipment*
FNSE	Framework for a Normative Science of Ethics
GIS	Geographic Information System
ICOMOS	International Council on Monuments and Sites
ISSS	International Space Syntax Symposium
ISUF	International Seminar on Urban Form
JCUD	Joint Centre for Urban Design
LPA	Local Planning Authority
NGO	Non-Governmental Organisation
OBU	Oxford Brookes University
OWEDC	Oxford West End Design Code
POS	*Plan d' Occupation des Sols*
SoADC	Stratford-upon-Avon District Council
SPG	Supplementary Planning Guidance
SSL	Space Syntax Limited
SST	Space Syntax Theory
UCE	University of Central England
UCL	University College London
UK	United Kingdom
UMRG	Urban Morphology Research Group
UNESCO	United Nations Educational, Scientific and Cultural Organization
US	United States
WEAAP	West End Action Area Plan
ZAC	*Zones d'Aménagement Concerté*

List of Figures

List of Tables

Chapter 1
The Relation Between Research and Practice

Vítor Oliveira

Abstract This introduction frames the debate for the chapters that follow. It opens the discussion by providing brief insights into the relation between science and practice in several different fields before addressing specifically the urban landscape and the cases of researcher-practitioners whose work has been fundamental in the development of urban morphology. It reflects on the main challenges, strengths, difficulties and threats, that the research-practice relation currently faces, highlighting the main contribution of the book, and of each chapter, to the debate.

Keyword Research and practice · Urban morphology · Urban planning · Urban design · Architecture · Researcher-practitioner

1.1 The Gap Between Science and Practice

'In theory, theory and practice are the same. In practice, they are not' Albert Einstein

The gap between science and practice is neither new nor exclusive to the urban landscape fields. On the contrary, across the span of social, physical and life sciences, the gap has been widely debated. This can be illustrated from different subject areas. Health (psychology), education, law and economics (business, management and accounting) are touched on in this introductory chapter, before focusing on research and practice in urban morphology.

In mental health and psychology, the origins of the debate on this gap (also labelled as the efficacy-effectiveness gap) can be traced back over the span of post-war years (Sobell 1996). In the mid-1980s Alan Ross concluded that behaviour therapy was at risk of losing its momentum due to an excessive preoccupation with theoretical developments and technical refinements. Against that background, he argued for a union that could facilitate a dialogue between researchers and clinical practitioners (Ross 1985). His call for bridging the gap was taken forward by Linda Sobell. She considered that the recognition of the gap and the implementation of a bridging

V. Oliveira (✉)
Universidade do Porto, Porto, Portugal
e-mail: vitorm@fe.up.pt

V. Oliveira (ed.), *Morphological Research in Planning, Urban Design and Architecture*,
The Urban Book Series, https://doi.org/10.1007/978-3-030-66460-2_1

process were two separate issues, and that if scientists were to have an impact on clinical practice they would have to change, or to use her words, they would have to learn a new way of doing business (Sobell 1996). She argued that the key to effective dissemination was to make practitioners true partners in the research, development and dissemination process.

While acknowledging advances in the integration of behaviour theory into clinical settings, in an overview paper published 20 years later, Shannon Stirman and her colleagues, stated that the majority of clinics were not using research evidence to inform care and that evidence-based treatments were not being delivered in routine care settings (Stirman et al. 2016). These American authors have introduced an important division, distinguishing between: diffusion, as the passive spread of information; dissemination, as the targeted distribution of information and intervention materials to a scientific audience with the intent to spread knowledge and the associated evidence-based information; and implementation, as the use of strategies to adopt and integrate evidence-based interventions and change practice within specific settings. In addition, they have identified four major factors that can influence the success of implementation: the outer context (including mandates, advocacy and communication, and funding and finance); the inner context (comprising leadership, climate, culture, structure and staffing and readiness for change); the characteristics of the individual (including attributes and attitudes); and, finally, the characteristics of the innovation. Focusing on the second and third factors, Gyani et al. (2014) argued that therapists prefer to use their clinical experience rather than research findings to improve their practice, and that the organization in which therapists work has an impact on these attitudes. Finally, Kazdin (2017) adds another dimension to the debate, by exploring how research can be used to extend treatments going beyond clinical practice and reaching a large number of people in need of clinical care that does not receive medical services.

A second area deserving our attention is education. Ellen Lagemann traces the history of education research and the long-time gap between science and teaching. She describes the conviction, in the nineteenth century, that education lacked a scientific basis; and how in the early twentieth century, a number of researchers (using testing, school survey and child study) started to focus on the administration and business of schools (rather than pedagogy) leading to the development of a research approach that would embrace the entire field (Lagemann 2000).

While education researchers and practitioners agree on the existence of the gap and on the need to bridge it, opinions differ on the causes and on the measures to close it. For instance, whereas some argue for large-scale experiments strictly controlled by researchers, others argue for small-scale studies in which researchers and practitioners would work together on an equal basis (Broekkamp and Hout-Wolters 2007). Throughout the twentieth century, policymakers funded multiple initiatives and centres to support improved dissemination of research findings to practitioners for these to guide improvements in practice (Penuel et al. 2015).

In a paper published in the mid-2000s, Donald McIntyre argues that the gap is mainly a problem of two contrasting kinds of knowledge: the pedagogical knowledge needed by classroom teachers in their everyday work and the knowledge that educational research is equipped to provide (McIntyre 2005). Between the two extremes there is a continuum of different kinds of knowledge. The British author explores three possible ways of dealing with this knowledge problem: through a dialogue between the two ends of the knowledge continuum; by the use of research strategies specifically designed to inform practice; and, through a negotiation starting from the centre of the continuum, through the establishment of 'knowledge-creating schools'.

The literature review authored by Hein Broekkamp and Bernadette van Hout-Wolters in the mid-2000s and the paper written by Ruben Vanderline and Johan van Braak in 2010 focus on the participants in this debate— researchers, intermediaries, teachers, teacher trainers, school leaders and policymakers (Broekkamp and Hout-Wolters 2007; Vanderline and van Braak 2010). While the former encourage these participants to take a many sided perspective instead of mono-causal analyses, the latter propose the establishment of 'professional learning communities' and the promotion of a 'design-based research' model. Amanda Cooper and Samantha Shewchuk extend the focus on one type of participant— the intermediary organizations (Cooper and Shewchuk 2015). The Canadian authors discuss the role of these intermediaries as knowledge brokers (intermediaries that act in both the education and the health sectors). Cooper and Shewchuk describe how this knowledge brokering occurs, through interaction among two or more different groups or contexts in order to improve the broader education system.

The research-practice relation in law is usually discussed under the labels of gap studies and legal effectiveness. In addition, and somehow relating the fields of law and education, the effectiveness of law schools' education as preparation for professional practice has been widely discussed, in a debate with origins in the nineteenth century (Van Buskirk and Filliter 2020).

The origins of gap studies in the United States can be traced in the so-called legal realists of the 1930s, who have then identified a difference between the law of books and the law in action (Gould and Barclay 2012). This line of thought gained new momentum in the late 1960s and early 1970s, in a time of optimism about the capacity of law to fight social injustices and promote civil rights. Initially, these gap studies considered the implementation or impact of Supreme Court directives in the United States (US). In subsequent studies, sociolegal researchers investigated legislation or court actions that were considered ambiguous, and assessed empirical material against standards of justice (such as the rule of law, arbitrariness, legality or due process). Finally, in the 1980s, at a time of decline in gap studies, researchers analysed the circumstances that could advance or retard implementation and impact.

Another framework for debating the science-practice gap in law is legal effectiveness. Legal effectiveness begins by identifying the goals of legal policy and assesses its success or failure by comparing the goals with the results produced. Where the results do not match the goals, attention is given to the factors which might explain the gap between the books and action (Sarat 1985).

The debate on this gap assumes different perspectives and foci, from the agents involved to particular sub-fields of law. The book chapter by Jane Goodman-Delahunty and Dilip Das, published in the mid-2010s, exemplifies the former focusing on practice lawyers. These authors analyse the action of prosecutors and public defenders working in the public sector, and identify an excessive workload that leads to ineptitude, with detrimental outcomes for justice (Goodman-Delahunty and Das 2017). The text by Mirko Bagaric and colleagues exemplifies the latter, addressing the sub-field of sentencing law in the US (Bagaric et al. 2017). Sentencing is the area of law on which there is the greatest gap between knowledge and practice. The dominant practice in the last four decades has been framed by sentencing objectives, that contradict research, and that justify heavy penalties and mass incarceration. These authors argue for an effective reform that reviews the proportionality principle, the impact of race on sentencing and the discouragement of the use of private prisons (Bagaric et al. 2017).

Finally, we focus on economics, more particularly on business, management and accounting. We start by an integrative literature review, addressing management, applied psychology (somehow relating two fields, psychology and economics) and human resource development. Oleksandr Tkachenko and his colleagues review the research-practice gap in these three areas (Tkachenko et al. 2017). Informed by this review, they suggest a model of key components affecting the interplay between research and practice: the process and product of knowledge production, individual characteristics of researchers and practitioners, and institutional and organizational forces. The way universities, and in particular business schools, are organized has been a main topic in the debate between science and practice. One of the main reasons for this is that universities are frequently asked by government and citizens to account for themselves. Within universities, due to their specific nature, business schools are under a tighter scrutiny. Claire Gubbins and Denise Rousseau use a paper by Herbert Simon, winner of the Nobel Prize in Economics, to frame this debate on designing better business schools (Gubbins and Rousseau 2015; Rousseau 2012; Simon 1967). According to Simon's paper in the mid-1960s, faculty research should be generating scientific knowledge to improve the world and to guide managerial problem-solving. Developing this line of thought, Gubbins and Rousseau (2015) offer five proposals for better translating research into evidence-based management: undertake rigorous, high-quality research; provide summaries and synthesis in the form of meta-analyses and systematic reviews that identify what we know and what requires further research; conduct research on real-world problems; translate the results for practitioner readability in the abstract and implications for practice sections; translate the results into actionable knowledge, in tangible forms like tools, decision rubrics and action guides.

With a different focus, Basil Tucker and Stefan Schaltegger compare the research-practice gap in management accounting in two different countries, Australia and Germany (Tucker and Schaltegger 2016). They use the description of Brownson et al. (2006) of four stages for converting research to practice: i. discovery, as the creation of knowledge through rigorous research that provides the scientific foundation of a discipline; ii. translation, as the adaptation of generalized findings from the discovery stage into a form useful to target populations; iii. dissemination, as the transmission of translated research findings to end users; and, finally, iv. change, as altering organizational practices on the basis of evidence from scientific research.

Common to Australia and Germany is the perception that communication is a major barrier. In Australia, practitioners access to research is seen to be a principal obstacle. In Germany, the limited relevance of topics researched by academics is perceived to be a significant barrier to research informing practice.

In reflecting on the fields of health, education, law and economics, it is evident that the gap between science and practice is widespread. In addition to touching on the origins, development and main characteristics of the debate in these fields, proposals for bridging the gap have been considered: from the focus on the factors affecting implementation (Stirman et al. 2016), to the idea of different kinds of knowledge (McIntyre 2005) and a model of key components (Tkachenko et al. 2017).

1.2 A Century of Urban Morphology: Cases of Researcher-Practitioners

The foregoing reflections on research-practice relationships in a variety of fields provide the context for focusing on the links that can be discerned between morphological knowledge and action on the city. The work of a number of researchers-practitioners is now explored. In comparison with the areas of knowledge considered in the previous section, urban morphology is a new field—even if the oldest tradition in the study of urban form, the antecedents of the historico-geographical approach (described in detail by Whitehand 1981), is considered. However, an analysis of the main works prepared by geographers and historians in Central Europe in the last decade of the nineteenth century and in the first three decades of the twentieth century, reveals little concern for action on the physical form of cities. Urban morphology within a geographical and historical tradition was more focused on the rigorous and systematic description and explanation of the urban landscape. But the specific characteristics of an architectural and planning tradition did facilitate the introduction of prescriptions.

One of the seminal contributions to the latter perspective came from Gustavo Giovannoni. The Italian architect was, indeed, one of the first to develop systematic morphological thought and translate it into action in the urban landscape. *Vecchie città ed edilizia nuova* is one of the most important texts written by Giovannoni. It theorizes the relation between new buildings, or new urban areas, and the existing city (Giovannoni 1913). The ideas of the permanence of the town plan and of the plan as palimpsest, and the discussion on how to act on the historical urban landscape with a clear focus on the elements of urban form (streets, plots, special and ordinary buildings) are key characteristics of this work. Rome, and in particular certain of its parts, such as the *Quartieiri del Rinascimento*, have a central role in this publication. While this part of the city offered the opportunity for Giovannoni to illustrate how his research on the intervention in historical centres could be translated into architectural practice, the *Quartiere Garbatella* (which is not analysed in his book) would illustrate his view on the design of new areas and on how these should be related to the extant city. Giovannoni also contributed to the avoidance of the demolition of parts of the historical centre of Rome.

One of the most notable examples of relating research on the urban landscape to practice was the work of Saverio Muratori (a former student of Giovannoni). At the end of the 1950s, he had the opportunity to apply the findings of the *Studi per una operante storia urbana di Venezia* (Muratori 1959) in the planning competition for *Barene di San Giuliano*. The book closed a decade of architectural research, historical study and didactic experience. After a reflection on a number of theoretical and methodological issues, it analysed the city of Venice dividing it into eight main areas. One of the fundamental elements of this comprehensive study was a set of plans referring not only to the existing situation in Venice at the end of the 1950s but going back to the eleventh century. In addition to these plans of the city and its different *quartieri* (elements of urban form of crucial importance), the book includes a number of plans, elevations and sections of Venetian buildings types. Muratori argued that the initial settlement could be reconstructed as a rational archipelago of parishes. In face of this, the gothic Venice somehow expressed a crisis in planning while the Renaissance Venice was a remarkable synthesis of a continuous and polycentric city.

In the *San Giuliano* competition, Muratori applied the 'designing in stages' methodology (Cataldi 1998). Accordingly, he prepared, not one single final proposal but, as many proposals as there are stages of urban growth in the history of Venice: in this case, three stages and therefore three proposals. Each of the three proposals adopted the designation of *Estuario* (Estuary) and it was a structural reinterpretation of one historical period (see also Maretto 2013). Estuary I is an interpretation of the tenth and eleventh centuries, at a time when the dominant urban development layout was a square centrally located within a group of islands, with a clear predominance of waterways over land routes. It corresponded to a city structured in a number of neighbourhoods comprised of islands linked to one another and to the mainland by bridges, and constituting self-contained units laid out along both banks of the *San Giuliano* estuary. Each island nucleus was a residential unit. Estuary II was an interpretation of Gothic Venice, with an urban organization in a comb shape and with a balance between canals and vehicular axes laid out in parallel. A set of self-sufficient neighbourhoods comprised of peninsulas laid out around the lagoon basin with their axis converging was proposed. Each was to have about 10,000 inhabitants. The plan consisted of building units with courtyards orthogonal to their peninsula axes. It is composed of one single residential building type, with three storeys and an arcade ground floor. Finally, Estuary III, the winning proposal, was an interpretation of Renaissance Venice, with a predominance of vehicular axes over the canals, and with buildings occupying the boundaries along the canals and releasing the inner space for land routes. An estuarine city was proposed, laid out along two strips parallel to the two banks of the estuary and gradually opening towards the lagoon, with a view of Venice. A double pattern of canals formed two series of flanking islands, which linked the features of the two preceding designs: the island and the peninsular systems. Despite its high quality, the plan (and in particular Estuary III) was not implemented.

While Muratori was developing his study on Venice, the German geographer MRG Conzen was analysing a number of settlements around Newcastle upon Tyne, including Alnwick (Conzen 1960). After having emigrated to England, but before moving to what would later be the University of Newcastle upon Tyne, Conzen trained for a career in town and country planning and worked as a practicing planner in north-west England for four years. This planning experience added a prescriptive dimension to the rigorous descriptive and explanatory perspective that Conzen shared with his associates in the German-speaking countries. A number of texts that Conzen prepared in Newcastle inform the construction of a theory of urban landscape management. Whitehand (1981) highlights three of these. In 'A survey on Whitby', Conzen made a major contribution to an investigation to provide a basis for an integrated plan for Whitby. In 'Historical townscapes in Britain: a problem in applied geography' and 'Geography and townscape conservation', he employed townscape analysis in the search for a theoretical basis for townscape conservation. These studies developed the potential of the concept of morphological region for planning purposes (Conzen 1958, 1966, 1975). The ideas are described further in the next chapter of the present book. In 1965 Conzen was active in attempts to prevent the demolition of part of the historical centre of Alnwick.

The idea of urban landscape management as a careful balance between conservation and transformation would be developed later and made more explicit by Jeremy Whitehand. This would be carried out in his research, particularly on the concepts of fringe belt and morphological region, and in planning practice, notably on the plan for Barnt Green in the mid-2000s that was prepared with Susan Whitehand, and in the supervision of PhD theses (see Oliveira 2019 for more details).

At the end of the 1970s, after teaching in Genoa and Florence, Gianfranco Caniggia (who followed the morphological approaches of Giovannoni and, particularly, Muratori) published, together with Gian Luigi Maffei, *Composizione architettonica e tipologia edilizia*. This notable work is in two volumes (Caniggia and Maffei 1979, 1984). The first is on the interpretation of the basic buildings, the second is on their design. The book focuses on the common buildings that make our urban landscapes, classifies them in different types and organizes these types in a logical sequence—a typological process. The method has considerable potential both for research and practice. In the early 1980s the potential of the method for practice was demonstrated in the design of the *Costa degli Ometti* neighbourhood in Genoa. Caniggia designed the neighbourhood in a promontory position. The layout follows the logic of the spontaneous formation of traditional promontory settlements. The line of the hill crest becomes the main 'through route' of the neighbourhood, and it is connected by stairways to the other streets that follow the natural curves of the promontory. The neighbourhood plots replicate the geometry and size of historical Genovese plots—4.5–6 m in width and of varying depth. Finally, single-family buildings were designed following the traditional relationships between street and building, and between plot and building (Corsini 2009).

Still in the 1970s a new morphological approach, space syntax, started to be developed at University College London (UCL). The first influential book of this approach,

'The social logic of space', was published in the mid-1980s by Bill Hillier and Julienne Hanson. It outlines a new theory, and innovative methods, for the investigation of the relation between society and space (Hillier and Hanson 1984). Five years after the publication of the book, 'Space Syntax Limited' (SSL), an architectural and planning practice company, was founded at UCL. Over the next three decades, the relation between research developed at UCL and the practice carried out at SSL would prove to be one of the most interesting processes (as opposed to singular or discontinuous links) of bridging knowledge and action on the urban landscape. One of the first emblematic projects of SSL was for the redesign of Trafalgar Square, London. By this time there had been the publication of two other important books (Hanson 1998; Hillier 1996), and the first International Space Syntax Symposium (ISSS) was taking place in the English capital. In the case of Trafalgar Square, the analysis of pedestrian activity patterns and the conception of a pedestrian movement model, has led to a design solution, developed with Norman Foster, that included a new central staircase, selective 'pedestrianisation' of the public realm and the re-connection of distinct square spaces.

In the early 1980s, Ivor Samuels started teaching at what would become Oxford Brookes University. One of the distinctive characteristics of this teaching experience was the involvement of students in a number of practice-oriented activities, including in France. A key concept that was developed was the urban tissue, as a particular combination of the main elements of urban form (Hayward and Samuels 2018). One of the most interesting places in which work was undertaken was *Asnières-sur-Oise*, with the collaboration of Karl Kropf, who, in his PhD thesis, would bring together, in the concept of urban tissue, the views of Conzen and Caniggia (Kropf 1993). Using the concepts of urban tissue and of levels of resolution, Samuels' team divided the urban form of *Asnières* into six levels of resolution which would be the bases for description, explanation and prescription. This means that the references for the new urban forms in each part of *Asnières* would be the existing forms. The approach involved a typo-morphological zoning, instead of the traditional functional separation. This lead to the identification of seven zones. For each zone a number of acceptable and unacceptable urban forms were illustrated. Five years after concluding the plan preparation, Samuels returned to *Asnières* for an assessment of plan implementation (a rather unusual, but crucial, procedure in planning practice – Oliveira and Pinho 2010). Based on a number of interviews of the main agents involved in this process, Samuels discussed a number of fundamental issues. These included the need to build a stronger political consensus to support a morphological approach, ensuring the presence of a qualified team for plan preparation and implementation, and recognition of the degree of control of design of detail adequate to each specific situation (Samuels and Pattacini 1997).

1.3 Morphological Research in Planning, Urban Design and Architecture

What is urban morphological research today? At the beginning of the third decade of the new millennium, urban morphology is a consolidated body of knowledge with a number of theories, concepts, methods and techniques to address the physical form of cities (Kropf 2017; Oliveira 2016). It can describe rigorously the elements of urban form— streets, street-blocks, plots and buildings— and their patterns of combination at distinct levels of resolution. It focuses on different urban landscapes, from historical kernels to peripheral areas, from planned to informal settlements. In addition, it can explain how these elements are shaped over time by different agents and processes of transformation. Finally, urban morphology can evaluate the impact of changes in urban form, framed not only by urban landscape criteria, but also by environmental, social and economic criteria.

And, what is practice in the urban landscape? Characterizing practice is more difficult than offering a picture of research, as it is more heterogeneous. There are major differences between planning and architectural practice. While planning takes place under legislative/political frameworks, addressing the fundamental dimensions of life in cities and aiming at prescribing the rules for their transformation, architecture occurs largely under a business framework, focusing on the design of buildings. Urban design is somewhere between these two activities. In most countries it is closer to urban planning, but in a few it can be closer to architecture, or it may not exist. A major consideration is the legal systems framing practice— from mandatory to discretionary systems, from systems closer to the 'urbanism' French tradition to those closer to the 'town planning' British tradition. Nevertheless, practice in relation to the urban landscape is, on the whole, well established. It includes a number of processes and procedures aiming at producing policies, plans and projects for the transformation or conservation of the physical form of cities. A continuous and detailed understanding of what practice is should involve investigation within that professional environment.

Table 1.1 summarizes the frameworks, objects, purposes, processes and results, that are prominent in the activity of researchers and practitioners (planners, urban designers and architects). It shows the fundamental differences between these two groups, and the existence of a gap between research and practice. This research-practice gap has been recognized on various occasions in the last two decades: for example, in editorials in the journal 'Urban Morphology' by Jeremy Whitehand (Whitehand 2000, 2003, 2007); in a special number of the journal 'Built Environment' edited by Stephan Marshall and Olgu Çalişkan (Çalişkan and Marshall 2011); and, particularly, in an ISUF Task Force on Research and Practice coordinated by Ivor Samuels (Samuels 2013). The latter has led to an analysis of how urban morphology is integrated in the curricula of practitioners (Ruiz-Apilanez et al. 2015) and how it is being used successfully in practice (Maretto and Scardigno 2016; Oliveira et al. 2014). It also led to the preparation of two manuals (Kropf 2017; Oliveira 2016). There has been a wide debate on theory and practice in viewpoints in 'Urban Morphology' (for example Barke 2013 and O'Connell 2013).

Table 1.1 Researchers and practitioners

	Urban morphologists	Planners and urban designers	Architects
Framework	Research	Legislative, political	Business
Object	Physical: streets, plots, buildings	Physical, environmental, social, economic	Physical: buildings
Purpose	Description, explanation, evaluation	Prescription, design	Design
Process	Individual, challenging knowledge Flexible Scientific language	Collective, establishing consensus Rigid, established routine Procedural and practical language	Individual, challenging style Rigid, established routine Practical language
Result	Theory, concept, method	Policy, plan	Project

What can urban morphology offer to professional practice? And how can practice be improved through the use of research? Urban morphology has developed a number of theories, concepts and methods that can describe and explain, with varying accuracy, the dynamics of urban form. This rigorous description and explanation is able to provide a number of recommendations for prescription and design. Furthermore, the impact of each proposed action on urban form, framed by a policy, plan or project, can be rigorously evaluated. Urban morphology offers planning practice knowledge of urban form that it tends to lack. The physical form of cities has been tending to lose prominence in planning, often being confused with land use. Research can offer practice a detached perspective on processes and procedures; a view with different time constraints from the routine of practice. Practice offers frameworks for thinking about urban form, including political, legislative and business aspects and awareness of wider settings in which the physical form of cities has to demonstrate its relevance. Simultaneously, it encourages research, most of which is undertaken by academics, to be practice-oriented.

Acknowledging the advantages, and defining the goal, of linking research and practice, how can we bridge this gap? First, it should be said that bridging the research-practice gap in the case of work concerned with the urban landscape is, as in many other fields, not a straightforward task. Largely in accord with Stirman et al. (2016), in their reflections on the research-practice gap in mental health, there are four types of difficulties: outer context, inner context, characteristics of the individual and characteristics of the innovation. Bearing this in mind, bridging the gap should be seen as a process. As Whitehand (2000) argues, occasional events, such as a guest lecture crossing the divide, a government planning official joining a research project, or the involvement of an academic in a development project, are not enough, as these occurrences are too infrequent to have much influence. There is a need to follow-up after the realization of each event to build an effective process over time. Depending on the socio-economic setting, the institutions, the individuals and the nature of the innovation, each bridging process can have its specific format: there is not a single form to this process. One fundamental purpose in the process should be the improvement of communication. As mentioned above, Stirman et al. (2016) distinguish three levels of communication: diffusion, as the passive spread of information; dissemination, as the targeted distribution of information and intervention materials to a scientific audience; and implementation, in the form of strategies to adopt and integrate evidence-based interventions and change practice patterns. Our focus should be on the third level. Using the framework of Brownson et al. (2006) the aim must be the change of practice. Improved communication between research and practice can lead to the establishment of common interests. The existence of radically different kinds of knowledge (McIntyre 2005) or contrasting positions of conceptual sophistication and practical relevance (Barke 2015) must be progressively deconstructed in favour of a continuum, in which a language shared by research and practice is progressively established.

1.4 The Researcher-Practitioner and the Structure of the Book

The researcher-practitioner, a figure who can have a fundamental role in the process of bridging the gap, is a principal focus of the book. The book brings together a number of policies, plans and buildings, developed in recent years in different geographical contexts and designed by academics who undertake research and practice on the urban landscape. The aim is not to promote one architectural style or one planning perspective, but to make explicit how a morphological view of cities can lead to the design of policies, plans and projects, and considering how they can be better suited to the urban landscapes to which they relate, and to the needs of the citizens living and working in these landscapes.

Most of the chapters present one policy, plan or project developed by one or more researcher-practitioners. The aim is to exemplify in detail the utilization of applied knowledge in a professional setting. Most chapters are in two parts. Most chapters present the morphological view of the author on the process of transformation or conservation of the urban landscape, and then exemplify how the underlying theory, concept or method is expressed in the urban landscape.

The book is in three parts. The first part addresses the link between urban morphology and the policies and plans that frame the tension between transformation and conservation of urban landscapes, including the dialogue between public and private interests. In Chapter 2, JWR Whitehand reviews the origins, developments and main characteristics of the historico-geographical approach to urban morphology. He shows the potential of this approach to planning practice in different geographical settings at widely differing scales. A similar approach is applied by Peter Larkham and Nick Morton in Stratford upon Avon. Chapter 3 describes how this morphological view has supported the definition of conservation areas and the design of

Supplementary Planning Guidance, offering an *ex-post* evaluation of these policies in the maintenance of town character.

The relation between the study of urban form and urban design is the theme of the second part of the book (comprising two chapters). Arguing for a more eclectic approach to urban morphology, Ivor Samuels exemplifies in Chapter 4 his view on the physical form of cities in two planning instruments, the *Plan d' Occupation des Sols* for *Asnières-sur-Oise* and the Oxford West End Design Code. He shows how the lessons learned in *Asnières* have been incorporated in Oxford, almost two decades later. In Chapter 5, Karl Kropf explores the possibility of deriving prescriptions for new development based on descriptions of existing and historic development, arguing for a normative science of the built environment. This framework is applied in a case study in Stratford-upon-Avon.

Finally, the last part of the book, consisting of six chapters, focuses on how urban morphology and building typology can frame the design and construction of new buildings, progressively moving from large to small scales of intervention. In Chapter 6, Wowo Ding shows how morphological research can be used to inform the revitalization of small communities in a country of rapid urban growth. In particular, she exemplifies how modern facilities can be integrated in the existing fabric and how modest interventions can have a satisfactory impact on the life of residents. In Chapter 7, Giuseppe Strappa explores the idea of reading and design as one thing, and uses the process-typological approach (particularly the concepts of typological process, route, base building, special building, node and pole) as a framework for the development of the *Terni* cemetery extension. Another example of the application of the process-typological approach is provided by Matteo Ieva. In Chapter 8, he presents his view of the urban landscape, arguing for critical action, for proactive research that denies authorship and supports intentionality, conditioned by the acknowledgement of the constitutive phase of the urban fabric. This morphological view is translated into a project for a mixed-use building in *Canosa di Puglia.* With a similar approach, Gianpiero Moretti (Chapter 9) explores urban history as a key driver for contemporary architectural projects. The idea is illustrated with a residential building in Quebec City.

The two last chapters present the design of two houses of which the architect is the owner of the building. Vítor Oliveira and Claudia Monteiro explore the relation between morphological reading grounded in town-plan analysis and the design of a new house in a nineteenth-century street in Porto (Chapter 10). The authors explicitly address a fundamental question of the research-practice relation, namely what to conserve and what to change? In this case, the architectural project conserves the most structural elements of the street and its urban tissue. In the last chapter, Frederico de Holanda, explores the relation between configurational analysis and the design of an atrium-house in *Sobradinho,* Brasilia. In addition, a critical appraisal of the house is presented, drawing on the owner's experience during two decades of utilization.

Acknowledgements The author is indebted to JWR Whitehand and Susan Whitehand for their detailed comments on an earlier version of this chapter. He is also grateful to Ivor Samuels for the reading of a draft version.

References

Bagaric M, Fischer N, Wolf G (2017) Bringing sentencing into the 21st century: closing the gap between practice and knowledge by introducing expertise into sentencing law. Hofstra Law Rev 45:785–850
Barke M (2013) Some thoughts on the first output of the ISUF Task Force on Research and Practice in Urban Morphology. Urban Morphol 17:134–135
Barke M (2015) Further thoughts on research and practice in urban morphology: a British perspective. Urban Morphol 19:96–99
Broekkamp H, Hout-Wolters B (2007) The gap between educational research and practice: a literature review, symposium, and questionnaire. Educ Res and Eval 13:203–220
Brownson R, Kreuter M, Arrington BB, True W (2006) Translating scientific discoveries into public health action: how can schools of public health move us forward? Public Health Rep 121:97–103
Çalişkan O, Marshall S (2011) Urban morphology and design: introduction. Built Environ 37:381–392
Caniggia G, Maffei GL (1979) Composizione architettonica e tipologia edilizia I: lettura dell'edilizia di base. Marsilio, Venice
Caniggia G, Maffei GL (1984) Composizione architettonica e tipologia edilizia II: il progetto nell'edilizia do base. Marsilio, Venice
Cataldi G (1998) Designing in stages. In: Petruccioli A (ed) Typological process and design theory. Aga Khan Program for Islamic Architecture, Cambridge, pp 159–177
Conzen MRG (1958) The growth and character of Whitby. In Dash GHJ (ed) A survey of Whitby and the surrounding area. Eton, pp 49–89
Conzen MRG (1960) Alnwick, Northumberland: a study in town-plan analysis. Institute of British Geographers Publication 27. George Philip, London
Conzen MRG (1966) Historical townscapes in Britain: a problem in applied geography. In: House JW (ed) Northern geographical essays in honour of GHJ Daysh. Newcastle upon Tyne, pp 56–78
Conzen MRG (1975) Geography and townscape conservation. In: Uhlig H, Lienau C (eds) Anglo-German symposium in applied geography. Giessen-Würzburg-München, 1973. Giessener Geographische Schriften 1975, pp 95–102
Cooper A, Shewchuk S (2015) Knowledge brokers in education: how intermediary organizations are bridging the gap between research, policy and practice internationally. Educ Policy Anal Arch 23:1–4
Corsini MG (2009) Tradition and modernity in the Costa degli Ometti neighbourhood in Genova. Edizioni Kappa, Roma
Giovannoni G (1913) Vecchie città ed edilizia nuova. Nuova Antologia, 1–50
Goodman-Delahunty J, Das DK (2017) Conclusion: closing the Gap between Law in Action and Research. In: Goodman-Delahunty J, Das DK (eds) Trends in legal advocacy: Interviews with prosecutors and criminal defense lawyers across the Globe. Routledge, New York
Gould JB, Barclay S (2012) Mind the gap: The place of gap studies in sociolegal scholarship. Ann Rev of Law Soc Sci 8:323–335
Gubbins C, Rousseau DM (2015) Embracing translational HRD research for evidence-based management: Let's talk about how to bridge the research-practice gap. Hum Resour Dev Q 26:109–125
Gyani A, Shafran R, Myles P, Rose S (2014) The gap between science and practice: how therapists make their clinical decisions. Behav Ther 45:199–211
Hanson J (1998) Decoding homes and houses. Cambridge University Press, Cambridge
Hayward R, Samuels I (2018) Moving urban morphology from the academy to the studio: the use of urban tissues in teaching and continuing professional development. In: Oliveira V (ed) Teaching urban morphology. Springer, Cham, pp 281–296
Hillier B (1996) Space is the machine. Cambridge University Press, Cambridge
Hillier B, Hanson J (1984) The social logic of space. Cambridge University Press, Cambridge

Kazdin AE (2017) Adressing the treatment gap: A key challenge for extending evidence-based psychosocial interventions. Behav Res Ther 88:7–18
Kropf KS (1993) An inquiry into the definition of built form in urban morphology. Dissertation, University of Birmingham
Kropf KS (2017) Handbook of urban morphology. Wiley, London
Lagemann EC (2000) An elusive science: The troubling history of education research. University of Chicago Press, Chicago
Maretto M (2013) Saverio Muratori: Towards a morphological school of urban design. Urban Morphol 17:93–106
Maretto M, Scardigno N (2016) Muratorian urban morphology: The walled city of Ahmedabad. Urban Morphol 20:18–33
McIntyre D (2005) Bridging the gap between research and practice. Camb J Educ 35:357–382
Muratori S (1959) Studi per una operante storia urbana di Venezia I. Palladio 3–4
O' Connell D, (2013) Morphology and design: The developing dialogue. Urban Morphol 17:52–53
Oliveira V (2016) Urban morphology. An introduction to the study of the physical form of cities, Springer, Cham
Oliveira V (ed) (2019) JWR Whitehand and the historico-geographical approach to urban morphology. Springer, Cham
Oliveira V, Pinho P (2010) Evaluation in urban planning: advances and prospects. J Plan Lit 24:343–361
Oliveira V, Silva M, Samuels I (2014) Urban morphological research and planning practice: a Portuguese assessment. Urban Morphol 18:23–39
Penuel W, Allen A, Coburn C, Farrell C (2015) Conceptualizing research-practice partnerships as joint work at boundaries. Journal of Education for Students Placed at Risk 20:182–197
Ross AO (1985) To form a more perfect union: It is time to stop standing still. Behav Ther 16:195–204
Rousseau DM (2012) Designing a better business school: CHANNELLING Herbert Simon, addressing the critics, and developing actionable knowledge for professionalizing managers. J Manag Stud 49:600–618
Ruiz-Apilánez B, Solís E, Ureña JM (2015) Urban morphological curricula in Spanish schools of architecture. Urban Morphol 19:146–156
Samuels I (2013) ISUF Task Force on Research and Practice in Urban Morphology: An interim report. Urban Morphol 17:40–43
Samuels I, Pattacini L (1997) From description to prescription: Reflections on the use of a morphological approach in design guidance. Urban Des Int 2:81–91
Sarat A (1985) Legal effectiveness and social studies of law: On the unfortunate persistence of a research tradition. Legal Stud Forum 9:23–31
Simon HA (1967) The business school: A problem in organizational design. J Manage Stud 4:1–16
Sobell LC (1996) Bridging the gap between scientists and practitioners: The challenge before us. Behav Ther 27:297–320
Stirman SW, Gutner CA, Langdon K, Graham JR (2016) Bridging the gap between research and practice in mental health services settings: An overview of developments in implementation theory and research. Behav Ther 47:920–936
Tkachenko O, Hahn HJ, Peterson SL (2017) Research-practice gap in applied fields: An integrative literature review. Hum Resour Dev Rev 16:235–262
Tucker BP, Schaltegger S (2016) Comparing the research-practice gap in management accounting: A view from professional accounting bodies in Australia and Germany. Account, Audit Account J 29:362–400
Van Buskirk K, Filliter G (2020) An examination of the effectiveness of Readers' Theatre as a teaching strategy in legal education. The Law Teacher 54:129–148
Vanderline R, van Braak J (2010) The gap between educational research and practice: Views of teachers, school leaders, intermediaries and researchers. Br Edu Res J 36:299–316

Whitehand JWR (1981) Background to the urban morphogenetic tradition. In: Whitehand JWR (ed) the urban landscape: Historical development and management; papers by MRG Conzen. Institute of British Geographers/ Academic Press, London
Whitehand JWR (2000) From explanation to prescription. Urban Morphol 4:1–2
Whitehand JWR (2007) Urban morphology and policy: Bridging the gap. Urban Morphol 11:79–80
Whitehand JWR (2003) Urban morphological research and practice. Urban Morphol 17:3–4

Part I
Urban Morphology and Planning

Chapter 2
Conzenian Research in Practice

J. W. R. Whitehand

Abstract The approach to the study of urban form developed by MRG Conzen gave rise to what was to become known as the Conzenian school of urban morphology. In recent decades it has provided a basis for applications in urban planning, especially conservation. The principal foundations of urban morphology were in the German-speaking countries of Central Europe. It was here that Conzen trained as a student of geography, history and philosophy before settling in England in 1933. After training as a town planner, he worked as a planning consultant, at the same time undertaking research in historical geography, before moving to a full-time academic appointment. Prominent among the concepts he developed were 'morphological regions', frequently termed 'urban landscape units' in recent usage, and the 'objectivation of the spirit', both of which drew on his German experience. But it was not until the last decade of the twentieth century that a sizable number of other urban morphologists began to explore his planning prescriptions, and most of them were working in Europe, where the International Seminar on Urban Form (ISUF) had been founded and the Urban Morphology Research Group (UMRG) in the University of Birmingham was expanding rapidly. The growth of urban morphology internationally was widespread after the turn of the century, notably in China and Portuguese-speaking countries. A major influence in China was the growth of the Urban Morphology Research Group in Peking University and its collaboration with the UMRG in the UK. Among the applications of Conzenian methods in urban conservation were those in the historic city of Pingyao, China and in an area adjacent to the Forbidden City in Beijing. Other World Heritage Listed Cities, or nominations for listing, in which the Conzenian approach was applied included St. Petersburg (Russia), Sibiu (Romania), Kuling (China) and Ouro Preto (Brazil).

Keywords Conzenian approach · German origins · International collaboration · Morphogenetic concepts · Prescriptive planning · Urban conservation

---

J. W. R. Whitehand (✉)
University of Birmingham, Birmingham, UK
e-mail: J.W.R.WHITEHAND@bham.ac.uk

V. Oliveira (ed.), *Morphological Research in Planning, Urban Design and Architecture*, The Urban Book Series, https://doi.org/10.1007/978-3-030-66460-2_2

Seeds of practice in most fields can be found in fundamental research. And urban morphology is no exception. Central to historico-geographical urban morphology is the Conzenian approach, and the ensuing discussion focuses on applications of this approach in urban planning, particularly by adherents of the Conzenian school. An appreciation of the origins of Conzenian urban morphology is fundamental to understanding these applications. It is therefore with the derivation and early development of this approach that this discussion begins, before pursuing at greater length usages and assessments of it in recent times in planning.

2.1 German Origins

Used increasingly over the three-quarters of a century since the Second World War, the approach to urban morphology developed by MRG Conzen has its roots in research in the German-speaking world in the nineteenth century. It began to take shape in Conzen's thinking in a geography thesis that he prepared as part of the Staatsexamen at the Friedrich-Wilhelms University of Berlin (now the Humboldt University). Submitted in 1932, this thesis recorded and interpreted the urban landscapes or townscapes of towns on the Havel River in an area extending some 90 km north and west of Berlin (Conzen 1932). Though the thesis was never published, excerpts from it were translated into English by Conzen's son, Michael P Conzen, after his father's death (Conzen 2004). This made widely available for the first time the beginnings of what was to become known by the 1980s as the Conzenian school of urban morphology.

Already in the Havel towns Conzen was distinguishing the different components of the townscape. In the town of Rathenow in particular he was characterizing the ground plan and house types and interpreting their distributional patterns in relation to the historical periods of their development. Here, as in his paper on the German colonization of East Prussia (Conzen 1945), his sensitivity to the historical layering in the landscape and his emphasis on the distinguishing physical characteristics of different historical periods were fundamental. They were precursors of his classic monograph on the English town of Alnwick, where he defined the townscape, or what was to become more widely referred to as the urban landscape, as 'a combination of town plan, pattern of building forms, and pattern of urban land use' (Conzen 1960, p. 3).

2.2 Urban Morphology and Planning

Conzen's perceptions on planning were already evident early in his career. Having trained as a planner in 1934–1936, soon after his emigration to England, and worked for a planning consultancy, 1936–1940, one of his earliest published papers was 'Towards a systematic approach in planning science: geoproscopy' (Conzen 1938).

However, the coming together of urban morphology and planning was not prominent in his published work until the post-war period. It remained implicit in his contributions to the *Scientific Survey of North-Eastern England* (Conzen 1949a, b).

Perhaps the earliest of Conzen's major publications in which the links between his morphological approach and planning were becoming evident was *Geographie und Landesplanung in England* (Conzen 1952). Here the connections are clear between building/planning legislation and physical forms on the ground. But it was the role of past planning in explaining landscapes that was considered rather than how existing landscapes provide a basis for 'prescriptive' planning. And it was the latter relationship that was to become an underpinning of Conzenian urban landscape management.

Arguably it was Conzen's contribution to *A Survey of Whitby and the Surrounding Area* that was the first of his publications in which urban morphology was presented as a basis for planning prescription (Conzen 1958). Fundamental to it were the plot-by-plot field surveys of building types and land use that he undertook of Whitby in 1956 and the remarkable detailed coloured maps for which these surveys provided the basis. Here were becoming evident his perceptions on the embodiments of history and geography in the urban landscape as foundations for preservation and architectural control. And this was before the urban conservation movement had gathered much momentum. Some of the bases that Conzen was setting out at this time for embodying future development in its historical roots can be traced back to earlier research in the German-speaking world: for example, by Freyer (1923), Granö (1929), and Spranger (1936).

2.3 Morphological Regions and Urban Conservation

The solutions to the challenges of conservation that Conzen was proposing in the 1960s (Conzen 1966) were not readily assimilated by English-speaking geographers and planners unfamiliar with Conzen's German-influenced ideas on geographical regions. Fundamental to his proposals was the superposition of delimitations of the three urban landscape components—town plan, building form and land use—to create a composite morphological regionalization. This procedure had antecedents in discussions among German-speaking geographers on the theory of regionalization. However, their discussions had not been specifically concerned with internal urban morphological divisions. The method that Conzen used was to derive a hierarchy of morphological divisions from the degree of boundary coincidence of the three basic form complexes within urban areas. This was the same in principle as the method employed by Granö (1929) on a much larger regional scale, but applied now by Conzen in urban areas it yielded the sharpness of boundaries inherent in the very nature of the urban landscape and its historical provenance.

For Conzen the three urban landscape components are linked hierarchically. The town plan 'contains', and provides a frame for, land and building utilization, which consists of land-use units that form separate parcels or plots which in turn are major influences on the building forms within them. The three urban landscape components differ in their response over time to changing functional needs. The town plan and to a lesser degree building form tend to reflect more markedly the 'past' pattern of land ownership and capital investment. Therefore, they provide a greater range and quantity of old established forms and thereby contribute more to the historical expressiveness of the urban landscape. Land and building utilization responds more rapidly to changing functional impulses. Its influence on the historical urban landscape is therefore more negative than that of the other two components (Conzen 1966, pp. 61–65).

Related to this perspective is the concept of the 'objectivation of the spirit' (Conzen 1966, p. 59). In geography this appears first in the work of Schwind (1951), but it had precursors in studies by philosophers—for example, by Freyer (1923) and Spranger (1936). Among the aspects that were important for Conzen was the embodiment in urban landscapes of not only the efforts and aspirations of their present occupiers but also those of their predecessors. This enabled people to take root in an area, acquiring a sense of the historical dimension of human experience. It helped to stimulate a less time bound and more integrated approach to problems and for Conzen was an important basis for urban planning, especially urban conservation. For him historical urban landscapes were assets to society both intellectually and as emotional experiences. These benefits of the 'objectivation of the spirit' were fundamental to a strategy for urban landscape conservation (Conzen 1966).

In the *Anglo-German Symposium in Applied Geography* held in Giessen, Würzburg and Munich in 1973, Conzen developed this perspective further (Conzen 1975). Aspects to which he drew attention included: (i) conservation of the physiognomic identities of urban areas and their constituent parts, entailing the establishment of spatial units of conservation; (ii) conservation of historicity and aesthetic quality as aspects of identity; (iii) conservation of the 'intelligibility' of the historical urban landscape in respect of orientation on the ground; (iv) preservation of human scale; (v) management of the building fabric by functional continuity, adaptation and concordant change; and (vi) conservation control of street spaces and street systems, and compatible traffic control. In fulfilling these principles the most important contribution of geographical analysis was identification of urban landscape units (or, what he termed at the time, townscape units (regions)) and the type and intensity of their historicity (Conzen 1975, pp. 85–86).

2.4 Research Concepts and Applications in Planning

Conzen's ideas on urban landscape conservation are intimately related to his ideas on the nature of the urban landscape. They are integrations of his artistic and historical interests and are only completely meaningful if viewed in the context of his morphogenetic concepts. His development of these was undergoing its most formative phases in the 1950s, 1960s and 1970s. One such concept that was to prove influential in both research and practice was that of the fringe belt. Like much of Conzen's thinking, the starting point for its development was in the German-speaking world (Louis 1936). It was deeply embedded in his fascination from his early days with the historical layering of urban development, in particular the unevenness with which urban areas extended over time. Employing detailed historical mapping he explored the way in which urban areas are made up of 'layers' of rapid growth, particularly in their residential areas, separated by zones of generally more extensive land use associated with planned limitations on urban growth or periods of reduction in residential development (Conzen 1960, pp. 58–65, 80–82). He showed in detail how the zones dividing up the urban area, often a succession of inner, intermediate and outer fringe belts, were part of the spatial pattern of urban landscape units that reflects the historical periodicities that characterize urban areas.

The earliest examples of planning proposals stemming from applications of such concepts in actual urban areas emerged from Conzen's own fieldwork. One such case was related to the fieldwork undertaken as part of the Survey of Whitby. This formed an important part of the basis for Conzen's explication of, and conservation recommendations for, the town. However, whereas in his study of Alnwick explication of the conceptual basis was fundamental, in Whitby the intended readership was more local, less specialized and more concerned with preservation and architectural control. Thus terms such as 'inner fringe belt' were avoided in favour of descriptions likely to be more meaningful to a wider readership, such as 'inner green belt' (Conzen 1958, p. 82).

2.5 Applications of the Conzenian Approach

Conzen himself published relatively little on the application of his ideas in relation to the management of specific urban landscapes. Summary recommendations for five mostly small British towns (Whithorn, Frodsham, Alnwick, Conway and Ludlow) for which he had undertaken field surveys between 1945 and 1964 were contained in his contribution to the *festschrift* for GHJ Daysh (House 1966), a volume that reflected Daysh's primary interest in applied research. In due course applications would be published by numerous academics and others working on urban morphology in various parts of the world, but only after a pronounced hiatus. It was to be a further two decades before applications of, and discussions of, Conzen's ideas on urban landscape management began to occur on a significant scale. This delay reflected in significant part the small number and limited availability, even within Britain and America, of the publications in which Conzen set out his ideas on urban landscape management.

In the meantime, a volume was published that sought to rectify this deficiency by reprinting and discussing four of Conzen's papers, including the two key ones dealing with applications in planning that had been first published in 1966 and 1975 (Whitehand 1981). However, it was not until the end of the 1980s and into the 1990s that publications on the Conzenian approach to urban landscape management gathered momentum (Larkham 1990; Whitehand 1989, 1990). This was a period of significant growth of the Urban Morphology Research Group (UMRG) in the University of Birmingham, several members of which were exploring links between Conzenian research and practice (for example, Barrett [1996], Jones [1991], Kropf [1993]). The momentum generated at this time in exploration of this relationship was a fillip to its further growth both within the UMRG and much more widely.

Conzen's own applications of his ideas on urban landscape management had all been in urban areas with long histories, substantial parts of them being traceable back to medieval times. In contrast one of the first applications by those following in his footsteps was in a predominantly suburban area, albeit containing a minority of traditional forms that were survivals, adapted to varying extent, from the rural landscape (Whitehand 1989). In this instance the mapping of urban landscape units commissioned by the owner of a potential development site (the plot of 'Old Hyrons' within the Hyrons Area unit, Fig. 2.1) was intended to form part of assessments relating to proposals for more intensive development. Fundamental from the Conzenian standpoint was compatibility of the proposed development within an existing urban landscape unit of traditional timber-framed buildings and neo-Tudor houses. However,

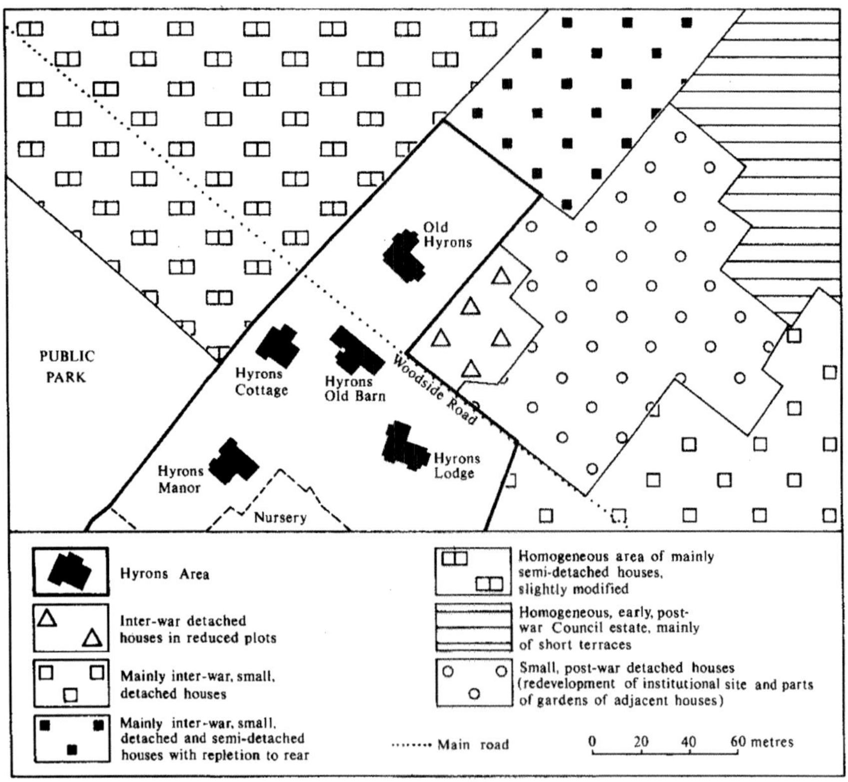

Fig. 2.1 Urban landscape units in part of east-central Amersham, UK in 1985 (*Source* Whitehand 1989, p. 13)

representations received by the local authority, most of them by residents in the vicinity, made scant reference to the characteristics of the areas recognized in the urban landscape units that had been identified. This was similarly the case with representations at a subsequent appeal to central government against the refusal of planning permission by the local authority.

Two contributions to a *festschrift* for Conzen published in 1990 reviewed his work on urban landscape management (still at this time perhaps more frequently termed 'townscape management'). Larkham (1990) remarked on the considerable potential of Conzen's mapping of town plan, building types and land use in the recognition and delimitation of morphological regions, the basis this provided for visual presentation of historicity, and the major extent to which conservation management in Britain fell short of the detail of Conzen's surveys (Larkham 1990, p. 354). He drew attention to the frequency with which the significance of underlying morphological features emphasized by Conzen, such as plot patterns, had been ignored by both developers and planning authorities (Larkham 1990, p. 364). He envisaged the future of urban conservation as ideally being an amalgamation of Conzen's detailed methods and ideas developed overseas, especially in Italy (Larkham 1990, p. 366; see also Samuels 1990, pp. 432–434). In the other contribution, Whitehand (1990) investigated four areas in south-east England that in the 1950s each consisted of predominantly large (0.1–0.5 ha) residential plots within 400 m of a town centre. Examination of the history of redevelopments in these areas in the 1960s, 1970s and 1980s revealed major differences between the Conzenian approach and actual planning practice. Change was far from being the outcome of a coherent strategy. Consciousness that the urban landscape was an amalgam of unitary areas was slight among all parties to the redevelopments. In each of the areas examined the motivation for change was unequivocally economic. The gulf between Conzenian precept and actual planning practice was large. Not only was the integrated view of development recommended by Conzen almost non-existent, but even if it had been adopted in the preparation of local plan documents it is hard to see how it could have survived the crude interactions that characterized decision-making.

Weaknesses in the conservation practice of local authorities in the UK when compared with the Conzenian approach were revealed by detailed studies by Barrett (1996) of conservation areas in the centres of Birmingham and Bristol. Central to Barrett's work was the mapping of plan units, building form units and land-use units, and the combination of these in urban landscape units (termed in this case townscape units), following the procedures set out by Conzen (1988) in his work on Ludlow. Figure 2.2 (reproduced from Whitehand 2009, p. 11) is based on the Birmingham part of her work. As is generally the case in Conzenian delimitations of urban landscape units, there was a hierarchy of units.

The foundation at this time of the International Seminar on Urban Form (ISUF) was an important development in both urban morphology generally and dissemination of the Conzenian approach. Starting as a small, almost entirely European, organization in 1994, it had by the end of the 1990s begun to have a much wider membership. This was facilitating the cross-disciplinary and wider international diffusion of Conzenian ideas as well as those of other schools of thought, notably deriving from

Fig. 2.2 Urban landscape units in part of central Birmingham, UK in 1970. Based on Barrett (1996) Figs. 4.16, 4.18, 4.23 and 4.25 (*Source* Whitehand 2009, p. 11)

the work of Saverio Muratori. In a published letter to MRG Conzen celebrating the approaching dawn of the new millennium, the first President of ISUF stressed the importance of interrelating scholarly endeavour and professional practice (Moudon 1999, p. 21).

However, it was not until the early years of the new century that commissioning of Conzenian surveys by governments began to occur. One of the first was for what had become by this time the lowest tier in the hierarchy of planning authorities in the UK—the parish. The parish council of Barnt Green, essentially a commuter settlement that had begun to develop in the mid-to-late nineteenth century in the vicinity of a railway station just south of Birmingham, sought advice from the UMRG on the preparation of its first Parish Plan. The initial mapping of urban landscape units, based on field surveys in 2005, was part of the public consultation for that plan. For that purpose, the term 'character areas' was substituted for 'urban landscape units' because it was felt that it would be more readily understood by the general public. For the same reason the term 'character areas' was retained in the Parish Plan that was published in 2006 (Fig. 2.3), and the term 'community spaces and utilities' was used instead of 'fringe belt', which would have been the correct term for a research readership. Among the factors accounting for the recognition of a hierarchy of areas ('subdivisions' in Fig. 2.3) were the presence of both purpose-built suburban roads and adapted rural lanes, and the fact that very large individual house plots had been subdivided to create culs-de-sac of higher density housing. Vegetation was taken into consideration in addition to the three components of urban form that Conzen took into account in his analysis of urban central areas. Assessment of applications for planning permission was to be undertaken in conjunction with a section of the Plan termed 'Planning for Character' which comprised explications and illustrations of each character area.

In a variety of other urban areas, landscape units of the type recognized by Conzen have been mapped for conservation areas and heritage protection areas for comparison with delimitations produced by local governments. Such a comparison was made by Bienstman (2007) in the Dutch city of Alkmaar. In this case an urban design consultancy was engaged by the City Council to prepare a plan of the 'spatial quality' of its Old Town. The intention was that the recognition of 'character areas' should guide the Council's planning decisions and provide a basis for enhancing the urban landscape. However, Bienstman (2007, pp. 195–196) noted that the criteria on which the Council's boundaries were based were not made clear and also revealed major differences between its boundaries and those of the urban landscape units based on Conzen's method (Fig. 2.4). At much the same time, though published several years later, a study in Stratford-upon-Avon, UK of the maps of character areas produced by the local planning authority, local residents and urban morphologists employing the Conzenian method, highlighted the weak assimilation into planning practice of relevant urban morphological research (Birkhamshaw and Whitehand 2012).

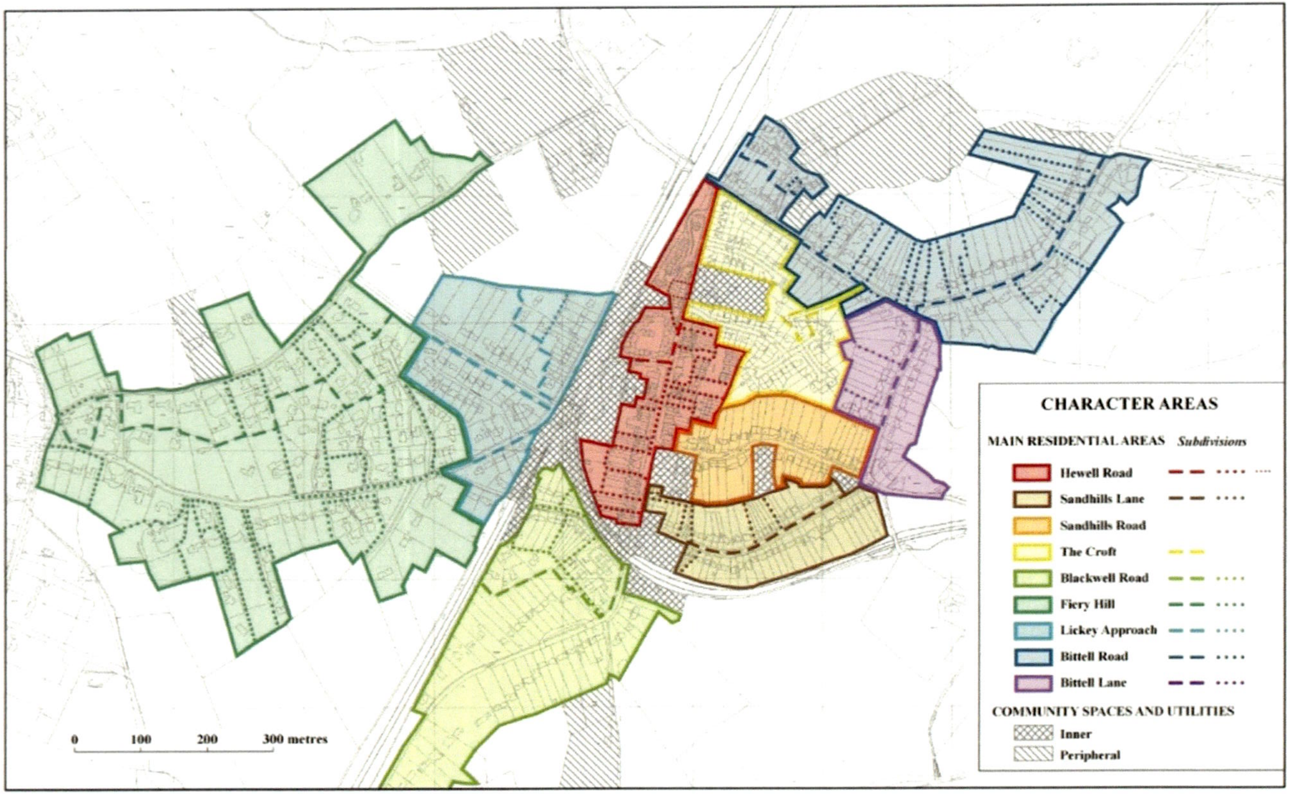

Fig. 2.3 The character areas of Barnt Green, near Birmingham, UK in 2005. Principal sources: field survey by the author and SM Whitehand, and Ordnance Survey Plans at the scale of 1:2500 (surveyed in 1885 and 1886; revised in 1902–1903, 1925 and 1969). © Crown copyright Ordnance Survey. All rights reserved (*Source* Whitehand 2009, p. 15)

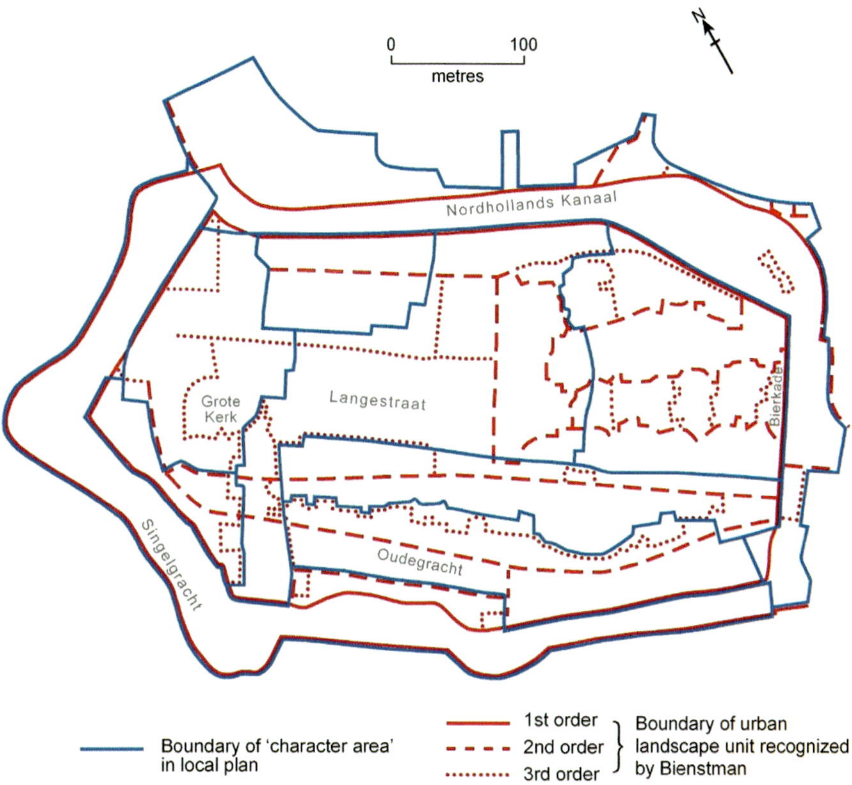

Fig. 2.4 Local authority character areas in 1999 (in blue) and urban landscape units recognized by Bienstman in 2005 (in red), in central Alkmaar, The Netherlands. Based on Bienstman (2007). Figures 5.11 and 6.2 (*Source* Whitehand 2009, p. 22)

2.6 International Collaboration

Until this time the Conzenian approach, both in historico-geographical research and in relation to planning, had been largely limited to Europe, although some investigations, mostly relatively minor, had occurred farther afield, including in Africa, where morphological research in Zambia had been supported by the Canadian government, North America and China, where work was in its early stages (Whitehand and Gu 2003). Now the interactions of researchers practically worldwide in some fields and the growing links between research and practice were associated with applications, not least in planning, over both a wider geographical span and involving more formal bodies, local, national and international. A notable example of this was the appointment in 2006 of the Head of the UMRG as Urban Planning Consultant to Pingyao County in Shanxi Province, China. The City of Pingyao had been included on the World Heritage List in 1997 and the mayor of the County was providing strong support for Conzenian morphological research, funded by the UK's Economic and

Social Research Council, that was underpinning that listing (Whitehand 2007). In particular, analysis of streets, plots and building block plans provided the basis for articulating the historical development of the city's exceptionally well-preserved layout, despite the shortcomings of the historical sources (Whitehand and Gu 2007a).

Notable features of the past decade in urban morphology more generally have included accelerated expansion practically worldwide. The growth of Regional Networks under the auspices of ISUF has been rapid. Particularly evident has been increasing activity in China and Portuguese-speaking countries. Expressions of the need for stronger links to planning have increased, not least among the growing number of adopters of the Conzenian approach.

Particularly significant in China has been the growth of an Urban Morphology Research Group in Peking University. A major influence on this development was the growth of links with the UMRG in the UK. An early manifestation of this relationship was the translation into Chinese by Feng Song of Conzen's study of Alnwick (Conzen 2011). This was in substantial part a product of a lengthy period of work in the UK, as were a number of subsequent publications on Chinese cities that employed, or were influenced by, a Conzenian approach (for example, Gu and Zhang 2014; Xiong et al. 2017; Zhang 2015). In a review of the origins and progress of urban morphology in China presented at the ISUF Conference in Valencia in 2017, a major part was devoted to the Conzenian school (Liu and Song 2017). Two years earlier Xiong (2015) was already suggesting that '*Alnwick* had become for some a kind of urban morphological bible'.

Cross-cultural comparisons have proved demanding, but in this respect the Conzenian approach has been shown to be readily adaptable. Successful collaborative ventures by workers from different cultures are well-illustrated in the comparison of the ground plans of Pingyao, China and Como, Italy (Whitehand et al. 2016). In both cities, planned developments over some 2000 years have left physical legacies that have provided frameworks that have yielded distinct patterns of present-day forms, notably relating to cosmological influences.

Developments in China that arguably were to prove more significant in the spread of the Conzenian approach in the Far East occurred at much the same time in Beijing. Redevelopments had burgeoned there since 1999, driven by the flourishing economy, but, despite China having become in 1985 a member of the UNESCO Convention Concerning the Protection of the World Cultural and Natural Heritage, destruction of its historical fabric had occurred on a massive scale (Dong 2006, pp. 195–196). Clearly formulated proposals for China's historico-cultural conservation areas were lacking. In this context Conzen's method of mapping the urban landscape as a basis for identifying conservation priorities was highly relevant to the historically important Zhishanmen area in Beijing. This small area, immediately north of the Forbidden City, lying just inside a large historico-cultural conservation area, had been selected by Beijing Municipal Planning Commission (BMPC) for special attention (Beijing Municipal Planning Commission 2002, 2004). However, the resulting conservation plan had little basis in research on urban form, as was soon to become evident.

Using Conzen's method, Whitehand and Gu (2007b) mapped urban landscape units in the Zhishanmen area and its immediate surroundings, based on a plot-by-plot survey in May 2006 combined with examination of past ground plans and other historical records (Fig. 2.5). In an area as small as this it is perhaps not surprising that the number of tiers in the hierarchy of units was only two. The key point, however, was the need to assess the future of existing forms in relation to the mosaic of developing character areas that the pattern of urban landscape units represented. This systematic integrated approach contrasted with the impressionistic one of the BMPC, in which individual sites and buildings were largely treated in isolation, divorced from the historical processes of which they were an integral part. There was a strong implication that the Conzenian approach, which had largely been limited

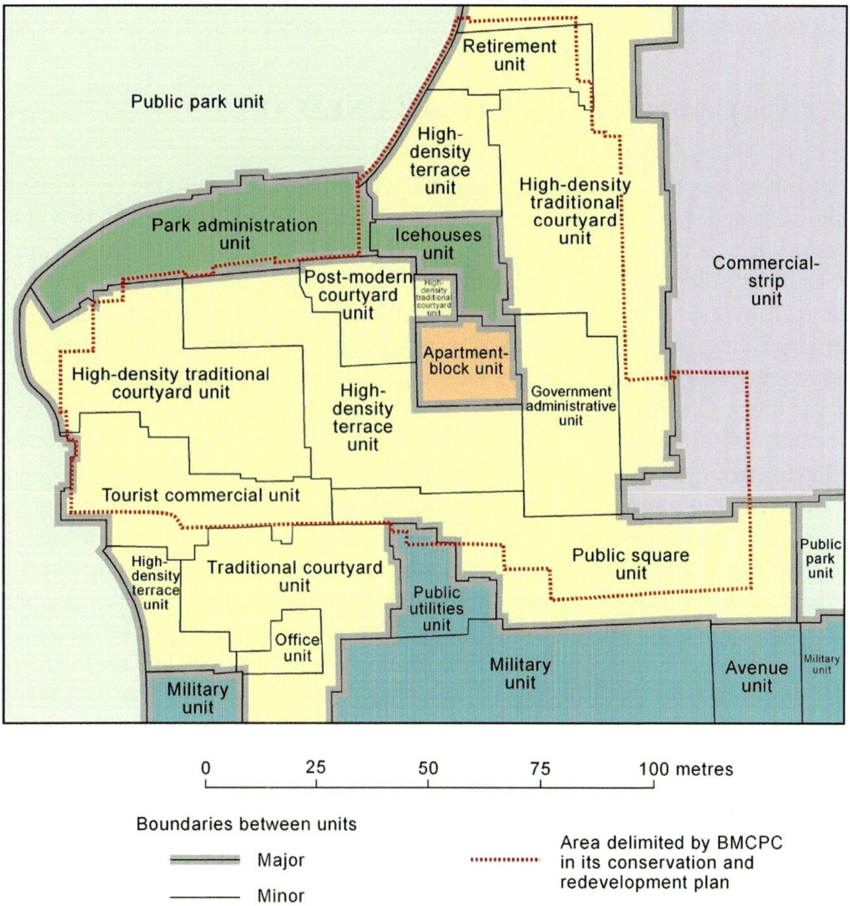

Fig. 2.5 Urban landscape units in the Zhishanmen area, Beijing, China in 2006. Based on Whitehand and Gu (2007b) Fig. 7 (*Source* Whitehand 2009, p. 20)

hitherto to its adoption in the West, was also appropriate in contrasting cultural conditions in the Far East.

A feature of the assimilation of the Conzenian approach in China has been the way in which aspects initially developed in the Western world have been followed up. The work undertaken in the Zhishanmen area that highlighted the contrast between the Conzenian approach and that adopted by the BMPC has been followed up by the much expanded Urban Morphology Research Group in Peking University (Peng et al. 2019). While recording several changes of land and building utilization and building type since the previous survey, 13 years earlier, they established that the plan units had remained unchanged. In addition to underlining the limitations of planning practice by local authorities, notably in conservation, the Group has contributed to the increasing evidence of mismatches between the Conzenian approach and procedures being applied by UNESCO.

2.7 The Conzenian Approach and UNESCO

Contemporary with the early stages of diffusion of Conzenian thinking into Eastern Asia, a change was occurring that had implications for its role more widely within conservation. This was the adoption by UNESCO of a World Heritage Cities Programme. As part of this programme, a Historic Urban Landscape initiative was launched in 2005 to raise awareness of the need to safeguard historic cities. The World Heritage Centre organized a series of meetings. At two of these, held in Jerusalem and St. Petersburg, urban morphological contributions were invited. It was argued by van Oers, Co-ordinator of the World Cities Programme, that among the most prominent of the issues emerging from the series of meetings was 'the importance of landscape', 'where all is layered and interrelated' (van Oers 2010, p. 12), and he drew attention to a move away from a rather static approach to the preservation of monuments towards consideration of dynamic processes (p. 13). This led in 2008 to the World Heritage Committee and the General Assembly of States Parties to the World Heritage Convention confirming their support for the 'historic urban landscape approach'. Inherent was the acceptance of change as part of the urban condition, although some contributors to the discussions found it difficult to accept this in relation to the core ideology of preserving monuments and sites as unchanged as possible (van Oers 2010, pp. 14–15).

At the World Heritage meeting in St. Petersburg in 2007, the urban morphological presentation focused on the Conzenian approach (Whitehand 2010). Most of the content was new to practically all those attending, hardly any of whom were familiar with the field of urban morphology. In illustrating the application of urban landscape units, one of the examples used was the delimitation of the World Heritage site of St. Petersburg itself. Figure 2.6 shows the boundary for that site proposed in 2005. It also shows the inner edge of the city's middle fringe belt in 2006, which was, and still is, a striking marker of the edge of the eighteenth- and nineteenth-century urban landscape for which this city has continued to be renowned. In contrast, the proposed boundary

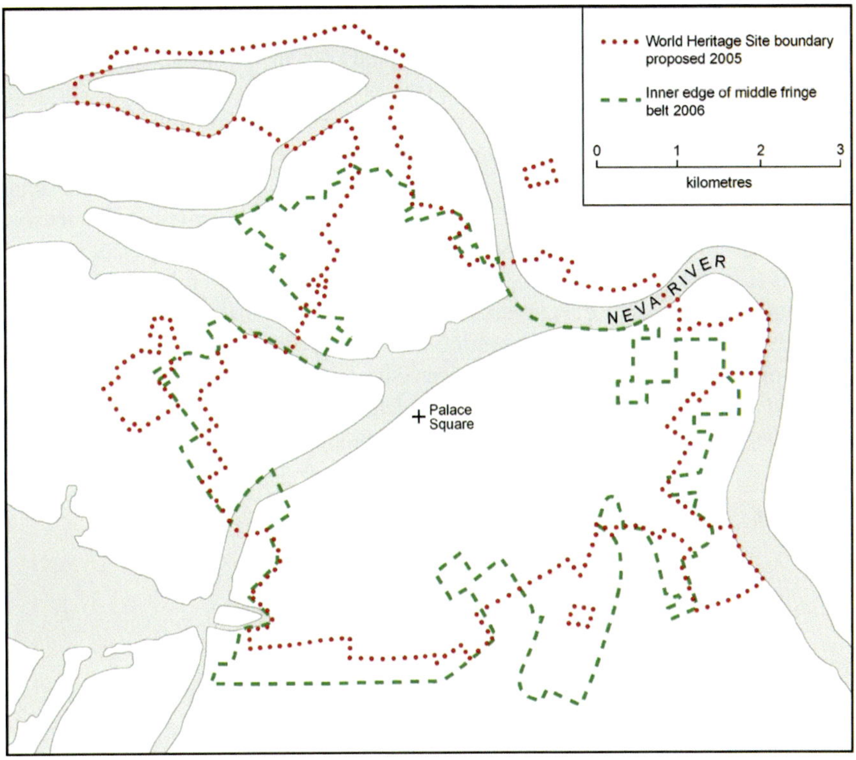

Fig. 2.6 World Heritage Site boundary proposed in 2005 (in red) and inner edge of the middle fringe belt in 2006 (in green), St. Petersburg, Russia. Principal sources: information provided by the Committee for the State Inspection and Protection of Historic Monuments, City of St. Petersburg; author's field survey; and Google Earth imagery (*Source* Whitehand 2009, p. 23)

of the World Heritage site excluded many areas inside that marker but included many outside it. It lacked a basis for heritage protection founded on the historico-geographical unity of what was being protected. However, this St. Petersburg example also illustrated that treating unity as if it were just a function of homogeneity was too simple. Some unified areas derive their unity from admixtures. For example, though fringe belts are unified by their role in the historico-geographical *grain* of the city, in some respects they are heterogeneous.

The presentation of the Conzenian approach in St. Petersburg had been very different in its perspective from other presentations at that meeting. Nevertheless, this did not deter the Romanian Ministry of Culture and Religious Affairs from issuing to the same author an invitation to take part in an Expert Workshop on improving its strategies for promoting the nomination of the historic centre of Sibiu for inscription on the World Heritage List. The resulting Conzenian presentation included the mapping of plan units in central Sibiu based on a field survey in 2009 in combination with a ground plan at the scale of 1:2500 that accompanied the nomination of the historic centre of Sibiu for inscription on the World Heritage List (Whitehand 2009, pp. 12–13). This mapping highlighted the way in which the recognition of types of Conzenian plan-unit boundaries (Fig. 2.7) articulate physical variations in historical development. For example, there is a contrast between the smaller, less regular units to the north-west (the Lower Town) and the larger, more regular units in the wealthier south-east (the Upper Town). This reflects the incorporation of small existing settlements in the Lower Town as it grew, and the paucity of extensive planned layouts (Fig. 2.8a). In contrast the Upper Town grew by a series of planned additions (Fig. 2.8b).

A notable aspect of the development of the Peking University Urban Morphology Research Group has been a collaboration with the UNESCO Beijing Office, which has responsibility for UNESCO in Eastern Asia. An example of this was the Forum on World Heritage Cultural Landscape held in Lushan, China in 2013, at which one of the key proposals was to 'develop a Statement of Intent which highlights

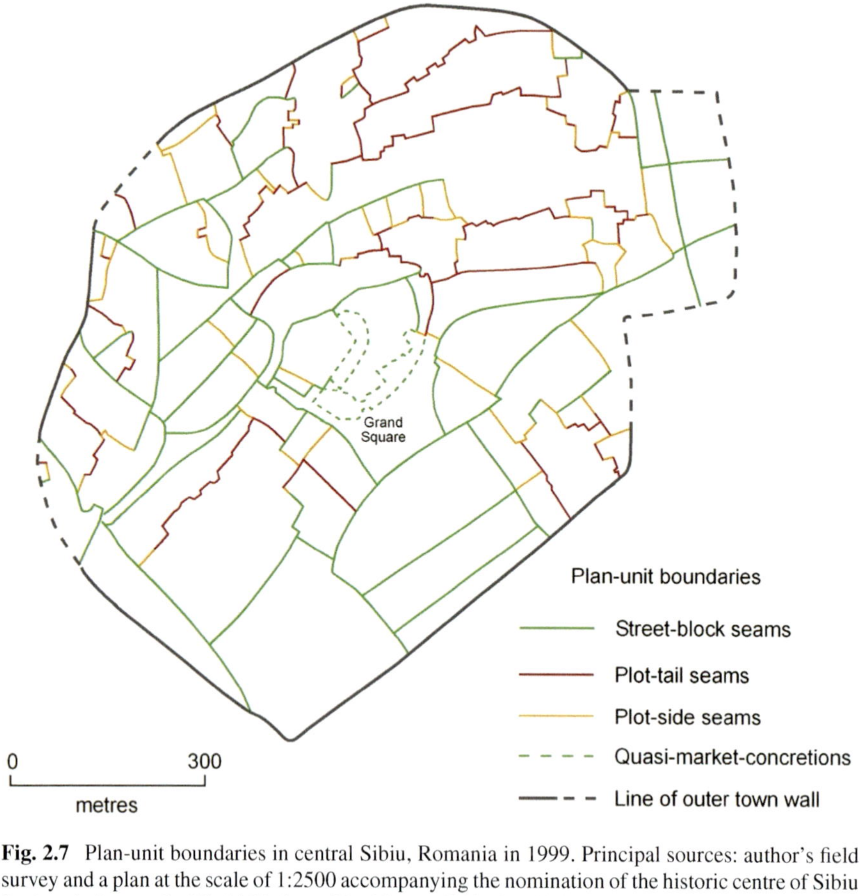

Fig. 2.7 Plan-unit boundaries in central Sibiu, Romania in 1999. Principal sources: author's field survey and a plan at the scale of 1:2500 accompanying the nomination of the historic centre of Sibiu for inscription on the World Heritage List (*Source* Whitehand 2009, p. 12)

Fig. 2.8 Plans of parts of central Sibiu, Romania in 1999. **a** Part of the Lower Town. **b** Part of the Upper Town. Based on a plan at the scale of 1:2500 accompanying the nomination of the historic centre of Sibiu for inscription on the World Heritage List (*Source* Whitehand 2009, p. 13)

the principles of managing World Heritage Cultural Landscape sites'. Contributions were sought not only from the Peking Group but also from the Head of the UMRG in the UK, who was 'invited to deliver a keynote speech on Cultural Landscape research as a basis for urban heritage management'.

An important aspect of the keynote presentation (Whitehand 2013) was consideration of the largely separate histories of cultural landscape research on the one hand and heritage management by UNESCO on the other (Fig. 2.9). Attention was also focused on the 'application' of cultural landscape research in the 'practice' of historical landscape management.

The discussion of the history of cultural landscape research from the nineteenth century to the present drew attention to the early contributions of geographers, historians and architects. It was papers by Schlüter (1899a, b) that had provided the basis for the study of urban cultural landscape as the organized field of knowledge that was to become known in the English-speaking world as 'urban morphology'. There was a fairly direct line from here to the Conzenian school and its application in cultural landscape management, particularly urban landscape management.

Quite different was the entry of UNESCO into the heritage aspects of cultural landscapes, both rural and urban. In 1964, the Second Congress of Architects and Specialists of Historic Buildings created the International Charter of Conservation

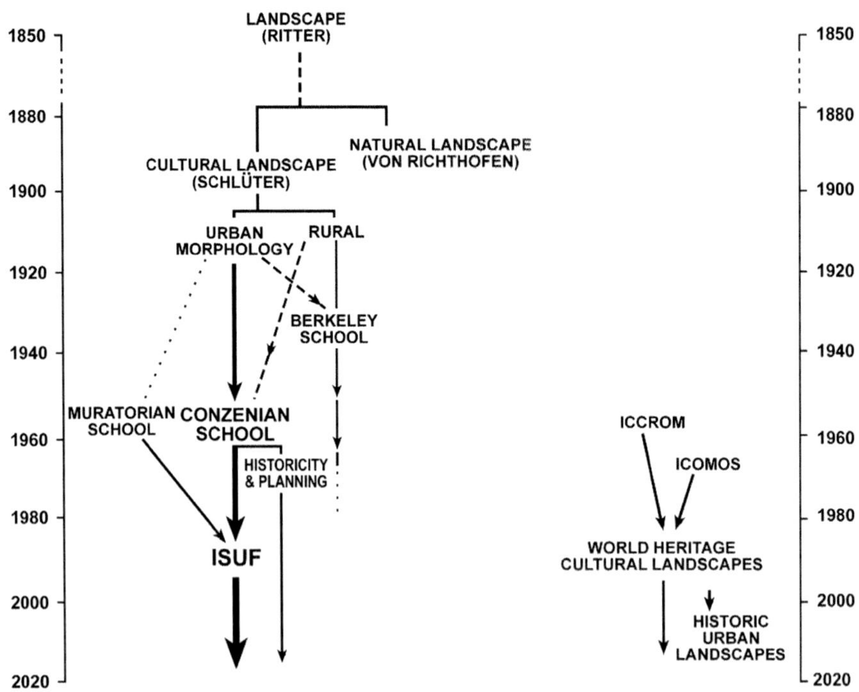

Fig. 2.9 The separate histories of cultural landscape research (left) and heritage management by UNESCO (right) (*Source* Whitehand 2013)

and Restoration of Monuments and Sites. The International Council on Monuments and Sites (ICOMOS) was put forward by UNESCO to implement this charter. However, the representation within ICOMOS of direct experience of research on cultural landscapes, including work on historical urban landscapes, was minimal.

The Lushan meeting, held in the town of Kuling, was timely, combining as it did both rural and urban landscapes in an environment in which direct exploration was possible literally within and around the site of the meeting. Hence it formed the basis for exemplifying aspects of the relationship between these two types of landscape. It should be seen in the context of earlier Conzenian analyses. One of these was of the World Heritage Site of Ouro Preto in Brazil (Conzen 2007). This dealt with a site not dissimilar to that of the town of Kuling within Lushan, in particular in its deeply dissected valleys. In this respect it provided an appropriate parallel, although the cultural context in which the forms were created was substantially different.

Kuling differs from practically all cities that had at the time of survey been subject to detailed morphological regionalization. It is not concentric-zoned in its layout. It has distinctive splays extending out from its core. These are heavily influenced by the topography. Rather than separating concentric zones, fringe belts separate splays. Furthermore, the occupied sites within the fringe belts are dispersed. They are referred to as 'dispersed fringe belts' in Fig. 2.10. The individual fringe-belt sites are associated with small areas of flatter land on otherwise very steep slopes.

In subsequent work on Kuling, the Peking University Urban Morphology Research Group drew attention to drawbacks of the UNESCO Historic Urban Landscape approach in the conservation of this type of heritage in China and the importance of taking into account academic research on urban landscapes (Xiong et al. 2017). In another paper, aspects of the protection of Kuling's urban landscape were explored in relation to a hierarchy of protection and management units (Xiao et al. 2017).

2.8 Reflections

Conzen's reputation is inseparable from his study of Alnwick, first published in English 60 years ago and much later in Chinese and Italian translations. This has a central place in the development of research in urban morphology. In the present discussion the focus has largely been on what has on the whole been the much later consideration of his research in planning, especially conservation.

The Conzenian approach, with its strongest roots in the discipline of geography, shares subject matter with a number of other fields of endeavour concerned with urban landscapes and their planning. The fact that these other fields have sometimes proceeded so differently reflects in particular their grounding in different languages and fields of knowledge. By the late-twentieth century, the Conzenian approach, with its origins in the German-speaking countries, was becoming evident in the English-speaking world of urban conservation. But other approaches to this aspect of planning had already achieved dominance in that world. The history of the field of urban morphology more generally was a factor, particularly in respect of the

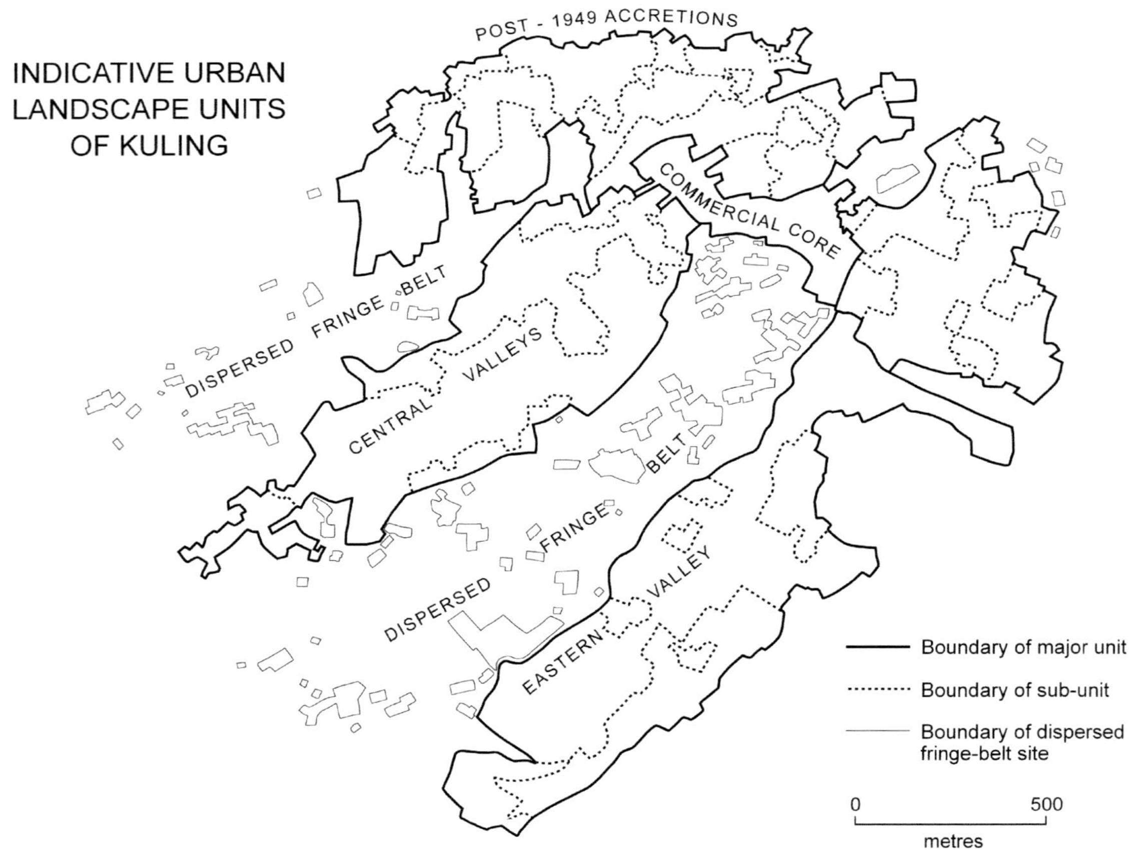

Fig. 2.10 Indicative urban landscape units of Kuling, Lushan, China. Based principally on a field survey by the author and Peking University Urban Morphology Research Group (*Source* Whitehand 2013)

disciplines within which it developed a footing. And this was also language related, with geography being on average less prominent in English-speaking countries than what perhaps seemed for some the more obviously urban-landscape related discipline of architecture.

This context is pertinent to the emerging concern in discussions about the relatively weak representation of morphological thinking in current planning debates. It has been particularly evident in ISUF and its house journal *Urban Morphology*. Hall (2013, p. 54) points out that the profound implications for planning of the observations made by MRG Conzen in his study of Ludlow (Conzen 1988) 'have never been fully and properly understood'. Similarly, McCormack (2013, pp. 45–46) cites fundamental misunderstanding of concepts such as the fringe belt by architects and urban designers who have had 'considerable influence on strategic thinking in the UK'.

In the last few years much attention has been drawn to the need for greater connection between the realities of practice and urban morphology as a field of research (see, for example, Song et al. 2016; Palaiologou 2017; Zertuche 2017). There has been sporadic reference to this in UNESCO meetings and publications (Jokilehto 2015), yet in the many contributions on UNESCO's historic urban landscape approach contained in *Reshaping Urban Conservation: The Historic Landscape Approach in Action* (Roders and Bandarin 2019) references to the Conzenian approach to urban landscapes are hard to find. There are many maps, and many more photographic views, in this volume of well over 500 pages, but maps and plans of the type that are fundamental to Conzenian historical landscape assessments are absent.

Van Oers (2015, p. 330) has drawn attention to the fact that 'urban morphological analysis proves to be an essential tool for the management of the historic urban landscape'. But he also notes that it is 'time-consuming'. In similar tone, Jokilehto (2015, p. 217) refers specifically to the Conzenian and Muratorian approaches as being 'fundamental for understanding the meaning and significance of the territories relevant to the definition of the Historic Urban Landscape', but adds that 'for many people such analyses may be too "scientific" to be easily digested'. Yet, if such research is indeed 'essential' and 'fundamental', the argument for applying it more generally merits a thorough assessment.

It has been shown in the present discussion of the Conzenian approach that in a variety of cultural regions it is both practicable and provides appropriate outcomes as bases for conservation. Nevertheless, the investment of effort in undertaking even some of the smaller-scale projects that have been considered is significant. In achieving the results discussed here in urban settlements large and small, and even in highly localized areas within them, reference has been made to various consultations and collaborations between researchers and practitioners. It could well be that more generally in urban morphology and kindred fields of planning, particularly related to conservation, there is a case for increasing such co-operation.

In tracing a few aspects of the lengthy international journey that the Conzenian approach has taken, both within research and from research into practice, no attempt has been made to draw parallels here with other fields of knowledge and practice. Possibly the closest to a parallel within urban morphology has been the Muratorian approach, which over a similar period also spread from an individual to an international environment. The two approaches have recently shown signs of interrelating (Maffei 2009), perhaps more than other urban morphological approaches. But therein is the stuff of another study.

Acknowledgements The author is indebted to Michael P. Conzen for his comments on a draft of this chapter.

References

Barrett HJ (1996) Townscape changes and local planning management in city conservation areas. Unpublished PhD thesis, University of Birmingham, UK

Beijing Municipal City Planning Commission (ed) (2002) Beijing Jiucheng 25 Pian Lishi Wenhua Baohuqu Baohu Guihua (Conservation planning of 25 historic areas in Beijing Old City). Yanshan Press, Beijing

Beijing Municipal Planning Commission (ed) (2004) Beijing Lishi Wenhua Mingcheng Beijing Huangcheng Baohu Guihua (Conservation plan for the historic city of Beijing and imperial city of Beijing). Zhongguo Jianzhu Gongye Press, Beijing

Bienstman H (2007) Morphological concepts and urban landscape management: the cases of Alkmaar and Bromsgrove. Unpublished PhD thesis, University of Birmingham, UK

Birkhamshaw AJ, Whitehand JWR (2012) Conzenian urban morphology and the character areas of planners and residents. Urban Des Int 17:4–17

Conzen MP (2007) Ouro Preto's urban morphology from a Conzenian perspective. Unpublished paper presented to the Fourteenth International Seminar on Urban Form, Ouro Preto, Brazil

Conzen MRG (1932) Die Havelstädte. Unpublished Staatsexamen thesis, Friedrich-Wilhelms University of Berlin

Conzen MRG (1938) Towards a systematic approach in planning science: geoproscopy. Town Plann Rev 18:1–26

Conzen MRG (1945) East Prussia: some aspects of its historical geography. Geography 30:1–10

Conzen MRG (1949a) Modern settlement. In: Isaac PCG, Allan REA (eds) Scientific survey of North-East England. British Association, Newcastle upon Tyne, pp 75–83

Conzen MRG (1949b) Geographical setting of Newcastle. In: Isaac PCG, Allan REA (eds) Scientific survey of North-East England. British Association, Newcastle upon Tyne, pp 191–197

Conzen MRG (1952) Geographie und Landesplanung in England. Colloquium Geographicum 2. Ferd. Dümmlers, Bonn

Conzen MRG (1958) The growth and character of Whitby. In: Daysh GHJ (ed) A survey of Whitby and the surrounding area. Shakespeare Head Press, Windsor, pp 49–89

Conzen MRG (1960) Alnwick, Northumberland: a study in town-plan analysis. Institute of British Geographers Publication 27, London

Conzen MRG (1966) Historical townscapes in Britain: a problem in applied geography. In: House JW (ed) Northern geographical essays in honour of GHJ Daysh. Department of Geography, University of Newcastle upon Tyne, Newcastle upon Tyne, pp 56–78

Conzen MRG (1975) Geography and townscape conservation. In: Uhlig H, Lienau C (eds) Anglo-German symposium in applied geography, Giessen-Würzburg-München, 1973. Giessener Geographische Schriften, Giessen, pp 95–102

Conzen MRG (1988) Morphogenesis, morphological regions and secular human agency in the historic townscape, as exemplified by Ludlow. In: Denecke D, Shaw G (eds) Urban historical geography: recent progress in Britain and Germany. Cambridge University Press, Cambridge, pp 253–272

Conzen, MRG (2004) The Havel towns: townscapes of the Havel region, exemplified by Rathenow. In: Conzen MP (ed) Thinking about urban form: papers on urban morphology, 1932–1998 (trans: Conzen MP). Peter Lang, Oxford, pp 83–100
Conzen MRG (2011) Chengzhen Pingmian Geju Fenxi: Nuosenboianjun Annike Anli Yanjiu (Alnwick, Northumberland: a study in town-plan analysis) (trans. Fong S, Zhongguo Jinazhu Gongye Chubanshe). China Architecture & Building Press, Beijing
Dong G (2006) Gudu Beijing 50 Nian Yanbianlu (A fifty-year evolutionary record of the ancient capital of Beijing). Southeast University Press, Nanjing
Freyer H (1923) Theorie des objektiven Geistes: eine Einleitung in die Kulturphilosophie. Teubner, Leipzig
Granö JG (1929) Reine Geographie: eine methodologische Studie beleuchtet mit Beispielen aus Finnland und Estland. Acta geographica Helsingfors 2:1–202
Gu K, Zhang J (2014) Cartographical sources for urban morphological research in China. Urban Morphol 18:5–21
Hall T (2013) The potential influence of urban morphology on planning practice. Urban Morphol 17:54–55
House JW (ed) (1966) Northern geographical essays in honour of GHJ Daysh. Department of Geography, University of Newcastle upon Tyne, Newcastle upon Tyne
Jokilehto J (2015) Evolution of the normative framework. In: Bandarin F, van Oers R (eds) Reconnecting the city: the historic urban landscape approach and the future of urban heritage. Wiley Blackwell, Chichester, pp 205–219
Jones AN (1991) The management of residential townscapes. Unpublished PhD thesis, University of Birmingham, UK
Kropf KS (1993) The definition of built form in urban morphology. Unpublished PhD thesis, University of Birmingham, UK
Larkham PJ (1990) Conservation and the management of historical townscapes. In: Slater TR (ed) The built form of Western cities: essays for MRG Conzen on the occasion of his eightieth birthday. Leicester University Press, Leicester, pp 349–369
Liu M, Song F (2017) Urban morphology in China: origins and progress. In: Proceedings of the twenty-fourth international seminar on urban form, Valencia, Spain, pp 749–758
Louis H (1936) Die geographische Gliederung von Gross-Berlin. In: Louis H, Penzer W (eds) Länderkundliche Forschung: Festschrift zur Vollendung des sechzigsten Lebensjahres Norbert Krebs. Engelhorn, Stuttgart, pp 146–171
Maffei GL (2009) The historico-geographical approach to urban form. Urban Morphol 13:133–135
McCormack A (2013) Informing and forming practice: the imperative of urban morphology. Urban Morphol 17:45–48
Moudon AV (1999) A letter from the President to MRG Conzen. Urban Morphol 3:21–25
Palaiologou G (2017) Urban morphology and world heritage practice. Urban Morphol 21:83–86
Peng D, Xiong X, Song F (2019) Tracking research on morphological evolution of the Zhishanmen area in Beijing: combining Conzenian approach with urban conservation methods. Unpublished paper presented to the twenty-sixth international seminar on urban form, Nicosia, Cyprus
Roders AP, Bandarin F (eds) (2019) Reshaping urban conservation: the historic urban landscape approach in action. Springer, Singapore
Samuels I (1990) Architectural practice and urban morphology. In: Slater TR (ed) The built form of Western cities: essays for MRG Conzen on the occasion of his eightieth birthday. Leicester University Press, Leicester, pp 415–435
Schlüter O (1899a) Uber den Grundriss der Städte. Zeitschrift der Gesellschaft für Erdkunde zu Berlin 34:446–462
Schlüter O (1899b) Bemerkungen zur Siedlungsgeographie. Geographische Zeitschrift 5:65–84
Schwind M (1951) Kulturlandschaft als objektivierter Geist. Deutsche Geographische Blätter 46:4–28
Song F, Dai Y, Li N (2016) Reconciling the theory of urban morphology and the practice of heritage conservation. Urban Morphol 20:159–161

Spranger E (1936) Probleme der Kulturmorphologie. Sonderausgabe aus dem Sitzungsbericht der Preussischen Akademie der Wissenschaften, Philosophisch-Historische Klasse, Berlin
van Oers R (2010) Managing cities and the historic urban landscape initiative: an introduction. World Herit Pap 27:7–17
van Oers R (2015) Conclusion: the way forward: an agenda for reconnecting the city. In: Bandarin F, van Oers R (eds) Reconnecting the city: the historic urban landscape approach and the future of urban heritage. Wiley Blackwell, Chichester, pp 317–332
Whitehand JWR (ed) (1981) The urban landscape: historical development and management: papers by MRG Conzen. Institute of British Geographers Special Publication 13. Academic Press, London
Whitehand JWR (1989) Residential development under restraint: a case study in London's rural-urban fringe. University of Birmingham School of Geography Occasional Publication 28
Whitehand JWR (1990) Townscape management: ideal and reality. In: Slater TR (ed) The built form of Western cities: essays for MRG Conzen on the occasion of his eightieth birthday. Leicester University Press, Leicester, pp 370–393
Whitehand JWR (2007) Historical urban landscapes and urban landscape management. Unpublished paper presented to a meeting of the officers of Pingyao County, Shanxi, China, April 2007
Whitehand JWR (2009) The structure of urban landscapes. Urban Morphol 13:5–27
Whitehand JWR (2010) Urban morphology and historical urban landscapes. World Herit Pap 27:35–43
Whitehand JWR (2013) Cultural landscape research as a basis for urban heritage management. Unpublished keynote presentation to the UNESCO Forum on World Heritage Cultural Landscapes in East Asia, Lushan, China, October 2013
Whitehand JWR, Gu K (2003) Chinese urban form: a European perspective. In: Petruccioli A, Stella M, Strappa G (eds) The planned city. In: Proceedings of the ninth international seminar on urban form. Uniongrafica Corcelli Editrice, Bari, pp 731–736
Whitehand JWR, Gu K (2007a) Extending the compass of plan analysis: a Chinese exploration. Urban Morphol 11:91–109
Whitehand JWR, Gu K (2007b) Urban conservation in China: historical development, current practice and morphological approach. Town Plann Rev 78:643–670
Whitehand JWR, Conzen MP, Gu K (2016) Plan analysis of historical cities: a Sino-European comparison. Urban Morphol 20:139–158
Xiao J, Shi C, Song F (2017) Utilization intensity assessment and conservative management of cultural landscape: an urban morphological approach, exemplified by Kuling town. Chin Landsc Archit 33(6):89–93
Xiong X (2015) An urban morphological bible? A view from China. Urban Morphol 19:99–100
Xiong X, Dai Y, Song F, Lu Y, Zhu D (2017) Morphology-based identification to OUV of HUL and its geographical process: the case of Lushan. Hum Geogr 32(3):36–43
Zertuche LN (2017) Practising the science of urban form. Urban Morphol 21:81–83
Zhang J (2015) Urban morphological processes in China: a Conzenian approach. Urban Morphol 19:35–56

Chapter 3
Conservation and (Sub)Urban Form: Reviewing Policy in Stratford Upon Avon, 2004–2019

Peter J. Larkham and Nick Morton

Abstract English local authorities have a legal duty to review their areas from time to time to consider whether conservation areas should be designated or the boundaries of existing areas changed. In the early 2000s Stratford upon Avon District Council fulfilled this duty by appointing external consultants to carry out these reviews, since the council had not only the 'heritage honeypot' of Stratford itself, but over 70 other conservation areas in the surrounding villages. Stratford had previously used urban morphological expertise to develop its district-wide design guidance, and it now appointed morphologists and planners to review the suburban areas adjoining the already-designated town centre conservation area. This chapter reviews the process of that study and identifies its limitations, but despite these limitations the report was accepted as Supplementary Planning Guidance, and the recommended conservation area extensions were designated. This chapter further reviews the extent to which the report's area-specific design guidance has been successful in protecting the character and appearance of these newly designated areas over the past 15 years.

Keywords Urban landscape management · Conservation areas · Designation · Area character · Planning policy · Supplementary planning guidance

3.1 Introduction

The interplay between urban morphology as an academic (sub)discipline, albeit supported by a wide range of more established disciplines ranging from geography, architecture and history to urban studies and urban design, and professional practices of urban planning, architecture and urban design, is complex and problematic. This chapter addresses one part of that problem using the example of a small-scale morphologically informed study of parts of a small English country town, albeit one

P. J. Larkham (✉) · N. Morton
Birmingham City University, Birmingham, UK
e-mail: Peter.Larkham@bcu.ac.uk

N. Morton
e-mail: Nick.Morton@bcu.ac.uk

V. Oliveira (ed.), *Morphological Research in Planning, Urban Design and Architecture*, The Urban Book Series, https://doi.org/10.1007/978-3-030-66460-2_3

with an international reputation and high tourist traffic. The study was commissioned from the authors and colleagues in 2004, adopted as policy in 2005, and this chapter evaluates its impact after 15 years in operation. The town is Stratford upon Avon in the English Midlands, and its numerous tourists demand a particular image and identity. In fact the town's contemporary character is commodified and increasingly aimed at, and created by, tourists; and this has major implications for the planning and conservation of the built environment (Hubbard and Lilley 2000). This has become most obvious in the past three or four decades, when a wider consensus about the importance of conservation has emerged (Pendlebury 2008) and processes of selection and management have identified an 'authorised heritage discourse' (Smith 2006). That discourse is very carefully managed and authorised in Stratford given the importance of its millions of tourists to the town's economy.

Part of the conservation discourse in England has been driven by the legal definition, in the 1967 Civic Amenities Act, of the 'conservation area'—areas of towns, identified normally by the local planning authority (LPA). These are 'of special architectural or historical interest, the character or appearance of which it is desirable to conserve or enhance' (1967 Civic Amenities Act, as amended). Hence 'character' and 'appearance' have become key concepts, elusive to define and subject to legal challenge over the years (Daniels et al. 1993; Skea 1996). This chapter explores the key questions of what gives a place its 'character', how can that be expressed in terms of its urban form characteristics, and how that knowledge might be effectively communicated in ways that relate to the needs of the English planning system. This is an application of urban morphology, and particularly that part of it derived from historico-geographical theories of urban form, to the professional context.

3.2 Conservation, Urban Morphology and Urban Planning

3.2.1 Tensions of Change and Non-change—Or Conservation

Our towns and cities are being continuously shaped and reshaped by economic forces, architectural tastes, planning ordinances, building controls, changing public fashions and a myriad of public-private regulations governing the form and use of space. Sometimes that change is fast, sometimes large-scale to the point where it can be thought of as catastrophic. At other times the change is small scale and incremental.

It is common to consider that cities are palimpsests of successive layers of redevelopment over time (Khirfan 2010; Martin 1968; Vâlceanu et al. 2014). Some features persist between layers: Conzen (1962) noted the longevity of street patterns in comparison to plot patterns and buildings, and Sabelberg (1983) studied the persistence of palazzi in Tuscan and Sicilian urban landscapes. Not only do persistent forms represent the investment and ethos of past societies, but it could be argued that retaining them contributes familiarity, character and identity; and contributes to sustainability through minimising resource use in replacing them. However, there are also studies of the speed and scale of change particularly in urban central areas, relating to issues such as changing land value and use (Hoyt 1933), and other constraints such as existing structures and societal values that might form a 'morphological frame' for subsequent development proposals (Larkham 1995). As While and Short (2010) suggest, the survival of particular structures, landscapes and forms reflects choices about what to retain in response to social, commercial and aesthetic opportunities, preferences and aspirations. These preferences are enmeshed in judgements about the value and meaning of different aspects of the past and the present (Graham et al. 2000). A significant thread of work on conservation policy has sought to make sense of factors that shape decisions about what to protect, why and how (i.e. what counts as valid 'heritage') especially at the local level. How is the familiar ordinary landscape differentiated from the special landscapes that should be protected? This has included work on the selectivity of statutory protection regimes and the role of advocacy groups in ensuring that certain aspects of the built form are preserved in the face of economic and political pressure for change (Hobson 2004; Larkham 1996).

Within the ebb and flow of urban change, it is, of course, important to note that the form and appearance of many cities are also shaped by powerful forces of conservation and preservation (although with changes between cultures and over time: Glendinning 2013; Jokilehto 1999). As a range of literature has demonstrated, the protection or promotion of 'heritage' involves the selection and (re)interpretation of certain aspects of the past to serve interests in the present: according to Graham et al. (2000, p. 17), 'heritage is that part of the past which we select in the present for contemporary purposes, be that economic, cultural, political or social'.

Local heritage narratives are unavoidably entrenched, to varying degrees, in the history and materiality of particular buildings and places (While 2007), and thus to the materiality of place-identity and character. These narratives are also consistently highly selective, prioritising certain histories, meanings, associations—and material artefacts—over others, and thereby favouring certain pasts, presents and futures (Atkinson et al. 2002; Massey 1995; Wright 1985). The promotional dimension of heritage management is especially marked given intensified competition between places for visitors and investment, with implications for identity, branding and promotion (Ashworth and Kavaratzis 2011; Uzzell 1996).

3.2.2 Character and Form

The Conzenian historico-geographical morphological approach considers a hierarchy of physical characteristics: streets (of all types) and their arrangement in networks; plots and their arrangement, bounded by streets, as street-blocks; and buildings. Land use is also a consideration as it has implications for physical form. These can be considered at scales from the micro-morphology of architectural and

Fig. 3.1 77 Tiddington Road, built in 1924: neo-Tudor possibly containing architectural salvage from Bradley Hall, Kingswinford (*Source* Photograph taken in 2005 by A Birkhamshaw)

structural elements of individual buildings, to the macro-morphology of settlement-scale patterns. They tend to change at different rates: street networks very slowly, buildings and especially their uses much more quickly.

In any one period, the dominant fashions and technologies create characteristic patterns, but the palimpsest and layering effect of different morphological periods over time and the differential survival of features provide, at any particular period, the character and form of the plot, street, district or settlement. These elements can, of course, be manipulated; with a Stratford example being the use of structural exposed timbers at various periods to create a town centre where the character and appearance are heavily influenced by 'Tudor' structures—genuine, replica and fake. This impact on character was recognised in the early 1920s (Abercrombie and Abercrombie 1923) and has enduring implications for conservation planning; and there are also examples in the interwar suburbs (Fig. 3.1 being the most prominent).

3.2.3 Planning and Conservation: Making It Work

English local planning authorities have a legal duty to review their areas 'from time to time' to consider whether conservation areas should be designated or the boundaries of existing areas changed—1990 Planning (Listed Buildings and Conservation Areas) Act. This duty has existed since the early 1970s–1971 Town and Country Planning Act, section 277(2), but this has never been undertaken frequently, systematically or consistently, with many authorities having insufficient resources for a regular review programme, instead responding in an ad hoc manner when particular issues arose (Daniels et al. 1993, p. 54). The reduction in resources and staff within authorities, and especially for conservation, since the 2008 economic downturn has exacerbated this problem (there has been a 36% reduction in conservation staff nationally between 2006 and 2017: Vines 2018).

The drawing of boundaries has been problematic, and approaches have varied since 1967. Some have been drawn widely, others extremely sharply focused on individual features. Some boundaries follow obvious landscape (morphological) features while others cut across them (for example medieval plot tails: Larkham 1990, Fig. 16.1). 'Obvious' features have been excluded on the instruction of the elected councillors who make the decisions (Birmingham Jewellery Quarter: information from former Conservation Officer). Areas can be designated because self-interested residents exert pressure on councillors (Larkham 2004, pp. 252–255).

Once designated and/or reviewed, the LPA has a duty under the same legislation both to produce enhancement plans, although this is rarely done; and to pay 'special attention' to planning proposals within the designated area. While it is difficult to prove in court that 'special attention' (as distinct from 'ordinary attention') has been paid, the development of area-specific policies, design guidance and related tools has been common; all helping to demonstrate that designated areas are treated differently in decision-making processes.

Conservation areas have become embedded in the English planning system (Pendlebury and Strange 2011): they are popular and numerous, but the system does have problems (Daniels et al. 1993, Skea 1996). A major issue is understanding the implications of small-scale change over extended periods (Skea 1996) since, even in conservation areas, the volume of development proposals can be high (Larkham 1996, Chapters 8–10).

3.3 The Stratford Example

Stratford upon Avon District Council (SoADC), responsible for a large and largely rural district with over 70 conservation areas (all village cores plus the tourist 'honeypot' of the town centre) hoped to review 10 areas per year by the early 1990s, although these reviews were being carried out by consultants after a tendering process (Daniels et al. 1993, p. 54). SoADC had previously used urban morphological expertise—employing Dr. Karl Kropf—to develop its district-wide design guidance (SoADC 2001, 2002). However Kropf had left SoADC by the time that this study was commissioned, and other staff have also subsequently left; so Kropf's suggestion (to the authors) that a focus on character areas and morphological underpinnings of the 2001 design guide might have been in their minds remains unproven. A more direct link is that the lead officer for this work was previously a member of staff of the School of Property, Planning and Construction at the University of Central England (UCE: now Birmingham City University).

3.3.1 Making the Policy: 2004

In 2004 there were public concerns about the effect of increased pressure for redevelopment, particularly on 'backland' plot tails, and their impact on the character of the residential approach roads to the town centre (SoADC 2005a, s2.1). Following its previous use of external consultants to carry out conservation area reviews, and with its recent history of morphologically informed policymaking, it now appointed morphologists and planners to review the suburban areas adjoining the already-designated town centre conservation area. An approach was made to the UCE School of Property, Planning and Construction. A positive response from David Chapman (DC: Professor and former Head of School) and Peter Larkham (PJL: Professor of Planning) secured a contract. As this was a short-term consultancy, Dr. Nick Morton (NM: Lecturer in Planning) joined the team, with Alex Birkhamshaw (AB: Ph.D. student, Birmingham University) as a research assistant. This was a very experienced team. DC was an architect and urban designer, formerly conservation officer with Birmingham City Council. PJL and NM had undertaken Ph.D.'s supervised by Jeremy Whitehand (founder of the Urban Morphology Research Group, UMRG, at the University of Birmingham), and had worked and published with him; PJL's research on conservation areas had included a report commissioned by the Royal Town Planning Institute (Daniels et al. 1993); AB was still a Ph.D. student supervised by Whitehand; his research was closely related to the subject of this consultancy and later resulted in a published paper (Birkhamshaw and Whitehand 2012).

The brief identified a series of 'mature residential areas'—a topic researched by several UMRG members (Larkham 1999a; Whitehand and Larkham 1991; Whitehand et al. 1992). These were the principal approach roads to the town (with the exception of the more commercialised Birmingham Road but including Alcester Road, Banbury Road, Evesham Road, Shipston Road and Tiddington Road), and three residential roads of varying dates and qualities (Avenue Road, Clopton Road and Loxley Road) (Fig. 3.2). The areas were tightly drawn on aerial photographs. Within each area we were to report on the distinctive character, and recommend whether some or all of the area was worth including within the town's existing conservation area designation.

The team carried out a search of relevant literature and policy; searched the planning history of the areas; carried out GIS-based measurements of key plot characteristics, including areas where changes were proposed or had been implemented; and carried out a plot-by-plot field and photographic survey, travelling daily from Birmingham for several weeks and basing themselves in the café of a major supermarket next to the railway station, adjoining one of the areas. Information was been obtained from a wide range of sources, including intensive on-site examination of areas and buildings; from historic maps in several local libraries and archives; from the local planning authority's records; from the County Sites and Monuments Record; and from local publications. For some areas, plot dimensions and other measurements were obtained from digital mapping. Building dates were obtained principally from visual inspection and comparison with historic maps.

Fig. 3.2 Examples of the study areas showing differing visual characters of mature residential areas: Banbury Road, top; Hathaway Lane (off Evesham Road), bottom (*Source* Photographs taken by P Larkham in 2020)

Based on the fieldwork, we attempted to persuade officers that some of the boundaries set for the area studies were illogical. An area boundary drawn along the centreline of a road will separate the two sides of the road; while these may have been developed at different times and in different forms, the road user perceives them as one unit. In particular, an early interwar social housing estate and a burial ground south of the Evesham Road clearly had distinctive characters, and social housing estates had been identified as under-designated in conservation terms (Daniels et al. 1993). However, the response was that there were no further resources available for this work, and any extension of the officer-delineated areas, while welcome, would have to be at the consultants' expense.

In the field it was felt that the most important 'headline indicators' of the condition of an area as it has changed over time were the conversion of original wooden windows to new materials (both aluminium and especially uPVC frames) and the conversion of front gardens to car-parking spaces (Fig. 3.3). Our recording of the latter was based on a field estimate of the amount of the front of the plot that has been converted. Conversion of timber to uPVC window frames was something of a conservation issue in the 1990s and 2000s (Larkham 1999b, pp. 364–365), and the increasing conversion of front gardens to hard-paved parking surfaces was not just a conservation problem, but was increasingly recognised as causing problems for drainage and biodiversity (DCLG 2010).

By 23 March 2005 a draft Residential Character Study report had been drawn up (Larkham et al. 2005). An important facet, informed by morphological work on town-plan analysis, was that many of these areas clearly consist of a range of sub-areas, each having distinct characteristics (Fig. 3.4). These are often the results of particular phases of design and construction. The report identified such areas and dealt with each separately, making specific development management policy recommendations (an example is given in Table 3.1) and suggesting whether or not the area character was sufficiently 'special' to be included within a conservation area; did not merit conservation area status but nevertheless had qualities that merited policy responses as an 'area of townscape interest'; or simply where generic policy responses would suffice (Table 3.2). It is worth highlighting that the management recommendations were deliberately focusing on a wider concept of urban landscape management, rather than solely on aspects of urban planning. They included comments on managing the

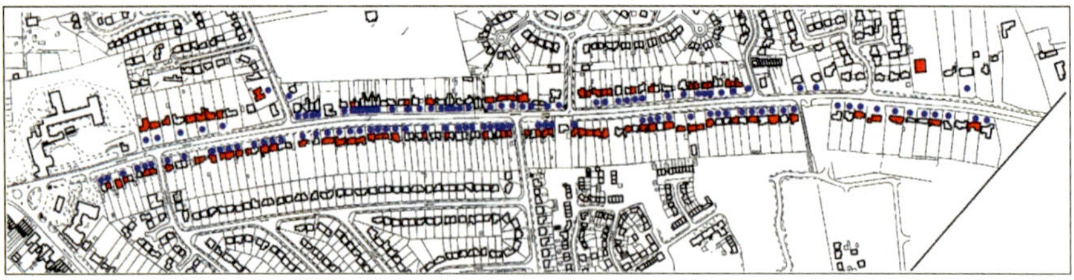

Fig. 3.3 Front garden conversions (blue) and replacement windows (red), Banbury Road (*Source* Larkham et al. 2005)

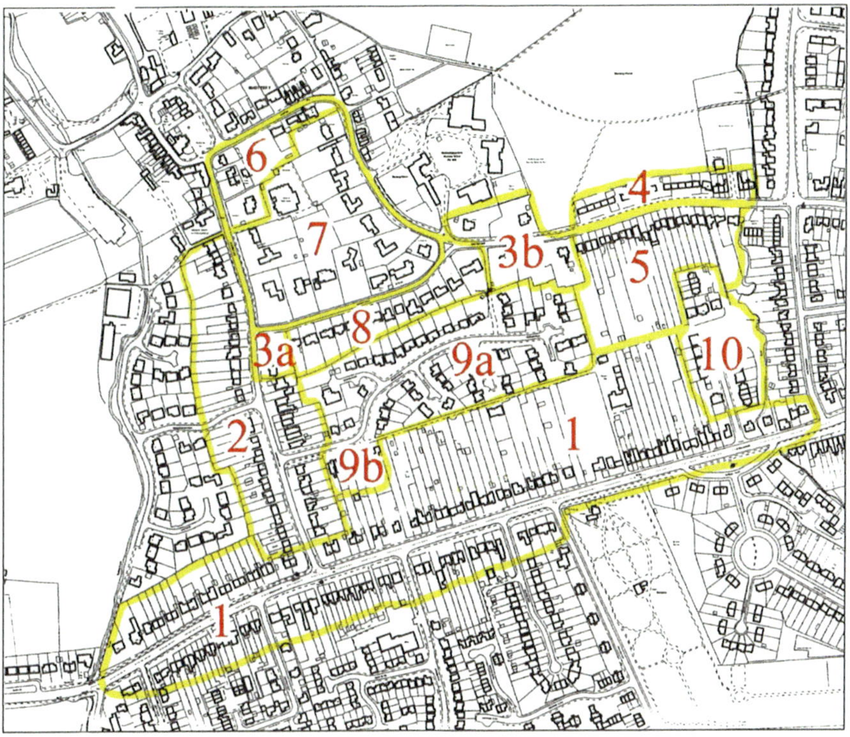

Fig. 3.4 The Evesham Road area, showing division into sub-areas each of identifiable and distinctive quality (*Source* Larkham et al. 2005)

public realm aimed at the local authority and public service providers, encouraging property owners to improve their properties, etc. This was highly aspirational, as the LPA was unlikely to be able to exert much influence on these points, but we felt that their inclusion in a public consultation document might stir some response.

It is worth examining the Evesham Road area to note the variety contained within some of the specified areas. While the character of the main approach roads to the town centre 'honeypot', which normally attracts over 2.5 million visitors each year (SoADC 2018a), was a key motivation, other 'backland' areas were included and some road frontages excluded (Table 3.2).

Table 3.1 Sample policy recommendations: the example of the Shipston Road area

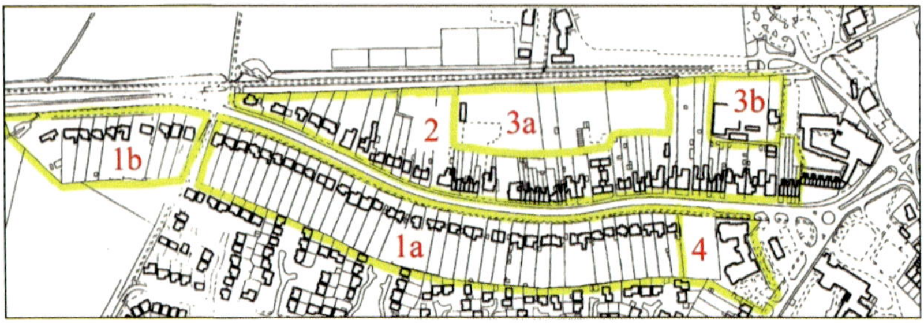

Areas 1a and 1b

The frontage development along Shipston Road is of varied form and character within a limited palette of typical inter-war styles. Further conversions of front gardens for car parking should retain as much of the existing front garden enclosures as possible. Roof conversions should have flat rooflights or appropriately-designed dormers on the front elevation. 'Terracisation' should be strongly discouraged. Individual properties (or pairs of semis) could be replaced with new development of appropriate scale and form, and ideally employing the dominant local materials of rendered upper floor and brick ground floor. Large-scale plot amalgamations (cf Area 4) should be resisted, and this would damage the characteristic of individual separate buildings. Any such new buildings should respect the local characteristic of building footprint being some 11-12% of the plot area on average. Building setbacks should be within the range of 9.4-13m.

Policy SUA.13 appears to apply only to the western side of the road; consideration should be given to whether it should also apply to the east in order to retain the domestic character, scale and use of these smaller inter-war houses.

The grass verge is a significant open area and it, and its trees, should be carefully maintained.

Area 2

In this already very varied area there is some scope for new building and conversions, particularly with some extension along the longer rear plots (subject to standard conditions on privacy etc). Plots are already at an average of 23% built over, significantly higher than Area 1. Building setbacks vary from 3-16m, so there is scope for variation around the average of 6m.

Nevertheless, plot amalgamations and large-scale apartment blocks or similar should be resisted.

If property owners can be encouraged to maintain and improve their buildings, the character and appearance of this side of the road could be improved significantly. There is particular potential in some of the larger Victorian properties to the north of the area.

Area 3

The precedent for significant development in Area 3 has been set, particularly by the butterfly farm. As this area is sheltered from easy view from the town and river, as many of the large plots were not originally directly connected with the road frontage plots, and as a number are neglected and in poor condition, it is suggested that there is scope for introducing some new development in this location. Good design would be vital, although there would be scope for high-quality contemporary design.

(**Area 4**, though shown as vacant on the base map, was under development in 2005 and so no recommendations were made for it.)

Table 3.2 Recommendations of 2005 report

Area	Recommend conservation area designation	Recommend designation of area of townscape interest	No designation recommended
Alcester Road		Yes	
Avenue Road	Yes		
Banbury Road		Yes	
Clopton Road		Yes	
Evesham Road	Small extension	Yes	
Loxley Road			Yes
Shipston Road		Yes	
Tiddington Road	Yes		

3.3.2 Adopting the Policy: 2005

The report was discussed by Stratford's Avon Area Community Committee on 24 March 2005. PJL presented the report and responded to elected members' questions. He highlighted the proposals for extending the conservation area boundary, and the suggestion that separate, smaller area designations would enable management policy to seem more focused. The overall reception of the study was very positive: 'members found these suggestions illuminating' and requested that officers arrange further training to allow them to be discussed in greater depth. The committee agreed to support the proposals as a basis for public consultation (SoADC 2005b).

A 4-week public consultation exercise was held in April and May 2005. Public responses were generally supportive (SoADC 2005a, s6.1), with one exception. A lawyer, resident in Tiddington Road, wrote to the Council threatening legal action on the basis that including his property within the conservation area, with the suggested policy guidelines, would breach his human rights because of the additional restrictions that would bring (SoADC 2005a, s3.9). The Council's response highlighted its statutory duty to review conservation area boundaries, and that any adopted policy would be in line with national guidance, and nothing more was heard.

Following the positive consultation response, officers repackaged the report and it was formally adopted as Supplementary Planning Guidance (SPG) on 25 July 2005 (SoADC 2005c). In then-current planning terms, SPGs were usually detailed documents, often guidance, supporting the approved Development Plan—in this case, the Stratford upon Avon Local Plan which was being produced at the time. An SPG document supporting an approved Development Plan, both of which had been through public consultation processes, carried considerable weight in the English planning system: it was a 'material consideration' in making decisions on whether or not to approve planning applications, for example.

Two small extensions were made to the existing town conservation area; the then Conservation Officer considering that extension was more effective (and, perhaps, faster) than designating the separate smaller areas recommended by the report (Fig. 3.5). The report's suggested small extension in Evesham Road was not accepted.

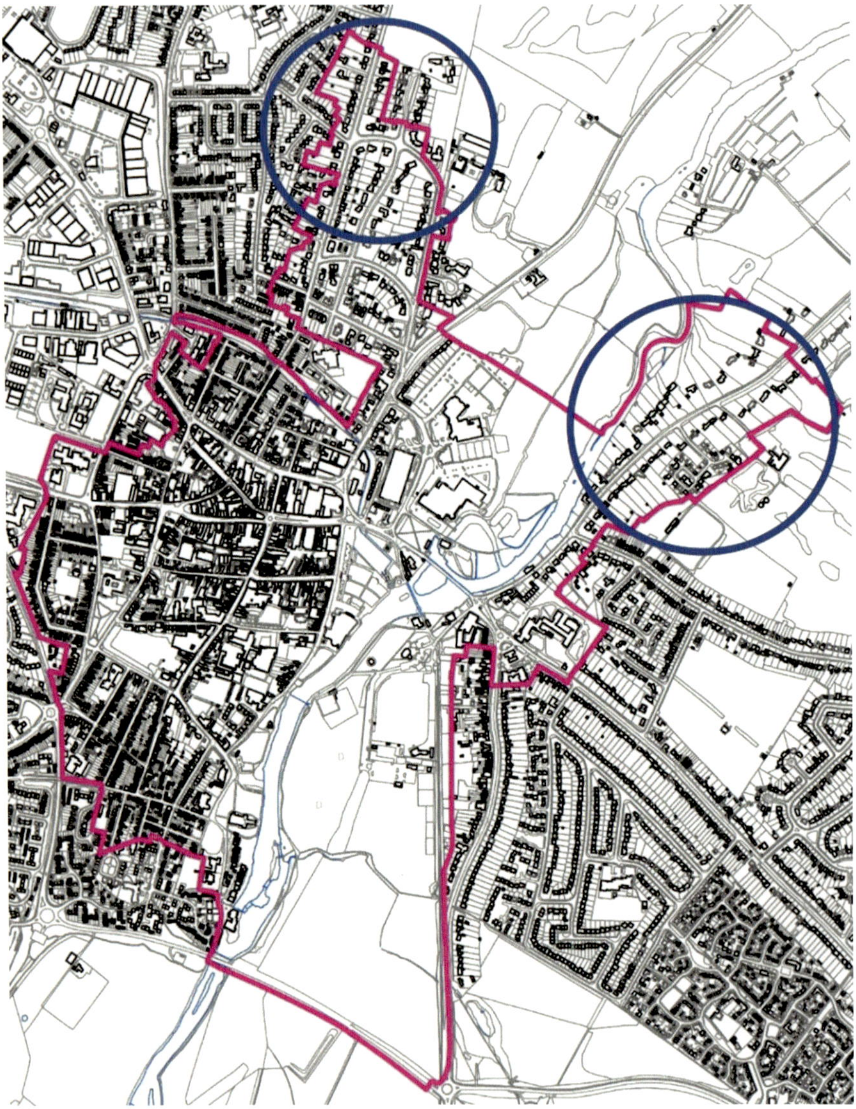

Fig. 3.5 The Stratford conservation area in 2020, with the approved extensions based on the 2005 report highlighted (*Source* Adapted from https://www.stratford.gov.uk/planning-building/conservation-areas-h-z.cfm)

3.3.3 Reviewing the Policy: 2020

In 2020 this book offered the opportunity to review the 2005 report and the supplementary planning guidance, to reflect both on the process of the research and on the impact of the policy. Inevitably, the political and legal contexts of planning had changed since 2005. In terms of planning law, the original survey coincided with the implementation of the 2004 Planning and Compulsory Purchase Act, which abolished Local Plans and Structure Plans, replacing them with Local Development Frameworks consisting of a series of policy documents ('local development documents'). The 2011 Localism Act tilted the ethos of planning firmly to a more local level (although in practice this has been limited, contradictory and problematic: Tait and Inch 2016). The National Planning Policy Framework (DCLG 2012, regularly revised) replaced previous, more detailed, policy guidance, and emphasised the importance of sustainable development: if a proposal could be demonstrated to be sustainable, there was a presumption in favour of granting planning permission. In fact, paragraph 7 redefined the purpose of the planning system: it is 'to contribute to the achievement of sustainable development'. However, 'character' and 'quality' are emphasised even outside designated conservation areas: paragraph 130 states that 'permission should be refused for development of poor design that fails to take the opportunities available for improving the character and quality of an area and the way it functions, taking into account any local design standards or style guides in plans or supplementary planning documents'.

More importantly for small-scale developments, the concept of 'permitted development'—normally specified types or sizes of development for which there is no need to apply for planning permission—was broadened significantly during this period, most recently by the 2015 Town and Country Planning (General Permitted Development) (England) Order. This was done 'to support growth in the economy' including the building industry, and among other changes it extended the maximum size for householder rear extensions (DCLG 2015). The explanatory memorandum makes no comments about the possible impact on area character or appearance. The continued extension of permitted development rights does have impacts on built environments and communities, and could be seen as undermining some of the 'public good' ethos of planning control (Clifford et al. 2019, Chapter 1).

The reforms introduced by the Labour government (1997–2010) intended to widen the remit of planning. However other governance processes and priorities, including corporatism, outsourcing (owing largely to the financial crisis) and managerialism reduced discretion at the local scale (Gunn and Vigar 2012). The reforms of subsequent coalition and Conservative administrations sought to support a market-led approach and to reduce central direction in favour of local discretion; although this clearly collided with the reality of retaining strong central direction and privileging developer interests (Allmendinger and Haughton 2019).

In Stratford, the Local Plan review was completed in 2006, and its documentation mentioned the 2005 report, although only in passing as it applied to only a small area of the District (SoADC 2006). The Core Strategy was adopted in 2016, containing policies on the historic environment and on design and distinctiveness. The Core Strategy is the key compulsory local development document, and every other local development document is built on its principles. It should be location specific rather than site specific (UK Government 2004). The Neighbourhood Development Plan was submitted in April 2017, subject to consultation and, at a referendum in November 2018, 91% of voters supported it. Formal approval was on 17 December 2018 (SoADC 2018b).

The updating of local planning documents following both national legislation and local policy changes meant that the 2005 Supplementary Planning Guidance became less directly relevant to the formalised and quasi-judicial process of making decisions on planning applications as subsequent documentation was published. The emerging Neighbourhood Development Plan had largely replaced specific mention of the SPG by 2014. As late as 2019 the SPG was still mentioned with other superseded documents in planning officer reports: common phrasing includes 'The [SPG] is a material planning consideration but, as it was adopted prior to the adoption of the Core Strategy and implementation of the NPPF [it] has limited weight in the current planning assessment' and 'While no longer having the status of a Supplementary Planning Document, it [the 2005 SPG] still contains substantial and relevant guidance on design' (see, for example, application 16/04057/FUL).

Nevertheless, the SPG, and other relevant design-related documentation, clearly influenced the wording of policies in the Core Strategy and Neighbourhood Development Plan, and were still being mentioned in 2019 by planning inspectors at appeal cases. In the English planning system, if a planning application is rejected by the LPA the applicant has the right of appeal to the Secretary of State. Planning appeals are managed by the Planning Inspectorate, and an independent adjudicator ('planning inspector') makes a recommendation to the Secretary of State. Core Strategy policies influenced by the SPG include Policy CS.9: 'all forms of development should improve the quality of the public realm and enhance the sense of place, reflecting the character and distinctiveness of the locality' and Policy CS.15 which seeks 'to ensure that the scale of development is appropriate to its surroundings, and the design should be related to and readily integrated into the existing form of the settlement, and should not have a harmful impact on the setting of the settlement'.

A more practical issue with reviewing the 2004–2005 study is the change of personnel (in both the local authority and university) and the loss of documentation. Moving house, office and even campus, with a reduction of storage space at each move, led to the only surviving documents being the draft report, the published SPG, and a file of the original photographic survey. The personnel changes in particular led this review to focus on documentary, rather than anecdotal, evidence.

3.3.3.1 Visual Inspection

The circumstances of early 2020, when this chapter was written, precluded repeating the detailed plot-by-plot inspection of the original survey, and instead Google Streetview images were reviewed to explore, as far as was possible, the extent of any further changes to the key features of windows and front gardens. The images were taken between mid-2019 and early 2020. There was little evidence of change in these features, under 5% per street. Some uPVC windows existing in 2004–2005 had been replaced by 'second generation' uPVC, some of which were in the currently-fashionable dark grey colour. This lack of change could be caused by the majority of desired changes having been completed by the date of the original survey; the worsened economic circumstances of much of the intervening periods having significantly reduced the pressure or the financial ability to make changes; or the effectiveness of the suggested policies.

Given the extent of changes in 2004–2005, in some streets in particular, the first possibility is most likely. The second is certainly possible, although the costs of these types of change could be relatively low. The third is possible, certainly in the years immediately after the acceptance of these policies and the surrounding publicity; although as most of these changes would have been 'permitted development', the local authority would have had no influence in these decisions.

3.3.3.2 Planning Applications and Appeals

A detailed study was undertaken of two of the streets for the full years 2006–2019. No applications for tree work, street telecommunications, etc., were reviewed. Tiddington Road was selected as part of this area was included in the extended conservation area, and the large plots and properties are some of the most desirable in the town—indeed in urban Warwickshire (Miller and Sandford 2020)—and so may have been subject to more development pressure than elsewhere. The number of applications, and the series of proposals for some sites, supports this suggestion. Evesham Road was reviewed as this was a more 'ordinary' area, with some of the plots fronting Evesham Road itself being particularly deep and thus potentially vulnerable to character-changing development, and the area character was complex resulting in the identification of numerous sub-areas in 2004–2005. The north side of Evesham Road (area 1), and areas 2, 3a, 7, 8, 9a and 9b (see Fig. 3.4) were studied.

A total of 129 applications (including appeals) were made for Tiddington Road, of which 72.1% were granted; and 113 for Evesham Road, of which 84.1% were granted (Tables 3.3 and 3.4). This can be compared to the national average, which has remained consistently between 87 and 89% between 2010 and 2019. Planning application statistics for England are compiled by the Department for Housing, Communities and Local Government (and its predecessors), from 2010 available at www.gov.uk/government/collections/planning-application-statistics. They were presented slightly differently pre-2010. They count planning appeals separately,

Table 3.3 Planning application data for Tiddington Road

Proposal	Result	2006	2007	2008	2009	2010	2011	2012	2013	2014	2015	2016	2017	2018	2019	Total
New build	Granted		1		3		1	1		1	2	1	4	1	2	17
	Refused		1	1							5	5		1		13
	Withdrawn											1	2	1		4
Side/front extension	Granted	4	1	3		3		1	1	2	2	1	3	1	1	23
	Refused							2	1				2			5
	Withdrawn			2												2
Rear extension	Granted	3	3	1	2		1	2	4	2	3		2	1	1	25
	Refused			1												1
	Withdrawn												1			1
Other minor work	Granted	2	2	4		1	2		1				1	1	1	15
	Refused	1						2			1					4
	Withdrawn			1								1			1	3
Variation of extant permission	Granted					1		1		1	2	1	2	3	2	13
	Refused	1														1
	Withdrawn						1					1				2

Notes
Appeals are counted separately (as separate processes with likelihood of new evidence)
Lawful development applications are not counted to minimise double counting
"Variation" includes procedural issues such as extension of time
If front and rear extensions are sought in the same application, it is entered here solely under 'front'

Table 3.4 Planning application data for Evesham Road sub-areas 1, 2, 7, 8, 9b (identified in Fig. 3.4)

Proposal	Result	2006	2007	2008	2009	2010	2011	2012	2013	2014	2015	2016	2017	2018	2019	Total
New build	Granted				1	1	1			1		1			1	16
	Refused					1				1		2		1		5
	Withdrawn											1	1			2
Side/front extension	Granted	1	5	3	4	2	1	1	3	1	1	2	1	2	4	31
	Refused							1								1
	Withdrawn		1				1					1		1		4
Rear extension	Granted	2	6	4	3[a]	1	3	1	2	2	2	2	3	6	1	38
	Refused										1					1
	Withdrawn	2		1					1							4
Other minor work	Granted	2				1		2			1		1			7
	Refused															0
	Withdrawn									1						1
Variation of extant permission	Granted				1	1	2	1	3	1	1	1	1	1	1	13
	Refused															0

Notes
Appeals are counted separately (as separate processes with likelihood of new evidence)
Lawful development applications are not counted to minimise double counting
"Variation" includes procedural issues such as extension of time
If front and rear extensions are sought in the same application, it is entered here solely under 'front'
[a]Includes one application for siting a mobile home to the rear of a house

whereas appeals are included in the Stratford analysis to give a better overview of decision-making.

There is more pressure for new development on Tiddington Road, with its larger plots and much higher-value property; and more for extensions in Evesham Road. Refusal rates higher than the national average are likely to reflect the more intense pressure for development in a desirable location—not just Tiddington Road but the whole town. The LPA's determination to manage development, to secure appropriate development in appropriate locations, is demonstrated by chains of applications including application rejected, appeal, appeal dismissed, application granted, then applications to vary the conditions of the planning permission. An example would be the chain of five applications and an appeal for the demolition of 78 Tiddington Road and construction of a replacement dwelling, between 2016 and 2019. Nevertheless, as is the case nationally, the vast majority of applications are granted. Even within the extended conservation area, it hardly seems as if the planning system acts to slow the development process, as has been a repeated political allegation throughout this period (for example Heath 2020).

There are still applications for infill development, although a suitable design on some of the larger plots would not damage the area character. One such plot subdivision on Banbury Road was approved with very little debate, and work on-site clearance is under way (as of mid-2020) (application 17/03020/FUL). The very large plots fronting Evesham Road are vulnerable to development: there is a rear alleyway access, and the area was even identified in the local development plan for possible development; but the intention was that it would be a larger unified scheme, and individual infill proposals have been successfully resisted. For example, application 14/01907/FUL proposed five houses on the tails of five plots fronting Evesham Road, with access along the rear alleyway. The application was refused, with the planning officer noting that the proposal did not relate well to the 'defining character' of the area as identified by the SPG. However in June 2020 one site was advertised for sale and the agent's board stated 'about half an acre (0.2 ha), potential development, poss[ible]strategic access strip' (Fig. 3.6). The SPG was not intended to stop development, but 'to establish general principles for residential areas in the town' and 'to seek an improvement to the design and setting of new development' (SoADC 2005a, s3.8). Purchase of one site as access would be risky.

The architectural styles along most of the streets studied vary considerably, although interwar (and, in the Avenue Road area, Edwardian) styles are dominant. Nevertheless the SPG noted that high-quality modern design might be appropriate for replacement dwellings in preference to historicist pastiche. Neighbours often disagreed, protesting that such a development 'would set a precedent for ultra-modern development which would result in severe loss of heritage' [along Tiddington Road] and 'ultra-modern design would not be sustainable' (application 16/00066/FUL). However this is not a convincing argument in such non-uniform areas; and the 2005 report and SPG prefigured paragraph 127 of the 2012 NPPF which stated that developments should be 'sympathetic to local character and history, including the surrounding built environment and landscape setting, while not preventing or discouraging appropriate innovation or change'.

Fig. 3.6 Site on Evesham Road advertised for sale in 2020 as possible access to facilitate large-scale backland development (*Source* Photograph by P Larkham in 2020)

Both application and appeal decisions for new development have specifically mentioned the SPG. In responding to application 08/00822/FUL within the extended conservation area on Tiddington Road, the conservation officer mentioned the SPG comments about ‘the importance of the varied scale and style of the buildings that prevents visual monotony’ and therefore objected to the proposed grouping of three similar houses. The appeal decision for application 16/02445/FUL, also within the extended conservation area, rested on only two factors: the impact of the new development on its neighbour, and on the character and appearance of Tiddington Road. In deciding on the latter factor the Inspector explicitly mentioned the SPG, finding that the proposed new building would ‘be dominant and visually intrusive within the streetscene [and] erode the gaps to the side of the plot. Therefore failing to reflect the character and distinctiveness of the locality’. Applicants have also used the SPG, as in the Design and Access Statement for application 14/03488/FUL, again within the extended conservation area, which quoted it extensively. One applicant even complained at appeal that the planning officer had not specifically referred to the Evesham Road character study, but the Inspector found that an unreasonable argument as ‘it is implicitly referred to by virtue of its adoption as part of the Design Guide’ (appeal relating to application 12/00891/FUL).

More common than wholly new construction is the substantial modification of existing houses. There are several examples of houses that have received very sizeable extensions, often accompanied by external insulation cladding and rendering. While this is partly driven by recent standards for insulation and energy costs, such changes usually occur when a property is purchased. The changes certainly affect the character and appearance of the buildings concerned especially as the small-scale architectural detail is often obscured and the building footprint often substantially increased. They are also often associated with changes to front gardens and property boundaries, although the example shown in Fig. 3.7 in 2005 and 2020 would still be categorised in the same way, although visibly very different (hedged boundary; garden with part driveway and part grass). Even these substantial changes to individual buildings have not been sufficient over 14 years to alter the overall character of these areas, since they remain dominated by individual houses on individual plots.

The online LPA planning documentation (accessible at https://apps.stratford.gov.uk/eplanning/) suggests that rear extensions, not visible from the main road frontage, are paid relatively little attention if it can be demonstrated that they do not block light from the neighbours' windows. Much more attention is paid to visible side and front extensions especially if the solid-void-solid pattern of buildings and spaces is disrupted. The joining-up of previously separate houses to form terraces has been identified in other studies of suburban form and conservation (Larkham 1999b, pp. 364–365; Whitehand and Carr 2001, pp. 173–175). There is relatively little evidence of pressure for this in Stratford, even in the areas of smaller semi-detached houses studied (Alcester Road). In part this may be because a number of such developments occurred before the commissioning of this research (Fig. 3.8). However, a number of properties have been extended so close to the property boundary that there is no space for maintenance access: while this is legal, it will cause long-term problems. English Heritage (since 2016, Historic England) did comment on several applications in Tiddington Road that 'to be more in keeping with the scale and nature of the development of the conservation area, the new building needs to occupy a smaller footprint' (for example application 16/00066/FUL).

Following wide debate, in 2008 the impermeable surfacing of gardens was removed from permitted development and so needed planning permission; but only one such application was submitted in these areas (15/00896/FUL, Evesham Road). This was approved with no objection or discussion. Several other applications sought dropped kerbs for vehicular access over the pavement, all of which were approved with no discussion of the motivation, i.e. for parking on the front garden, and its visual/environmental impact.

One of the 2005 report's recommendations was that, as some of the fields along approach roads into the town became developed, the dominant rural character should be protected by retaining the tall hedge lines. In some places this has been done successfully. However the continued policy of targeting new residential development in Stratford itself—3,500 new homes are planned between 2011 and 2031, of which some 2,400 had been built by 2018 (SoADC 2018b)—has direct and indirect character-changing impacts, as do the associated new retail facilities and large residential care homes being developed. The 'approach roads' covered in the 2005 report now no longer start at the urban edge: the town is expanding, and its overall character is changing.

Fig. 3.7 House on corner of Banbury Road and Dale Avenue, 2005 and 2020. Large extension and rendering, replacement of hedge, drive and grass (*Source* Photographs taken by A Birkhamshaw in 2005, top, and P Larkham in 2020, bottom)

Fig. 3.8 Pre-2005 extensions of semi-detached pair on Banbury Road (*Source* Photograph taken by A Birkhamshaw in 2005)

3.4 Conclusions

This review of the 2005 report demonstrates that, for more than a decade, it was clearly useful in local decision-making at the level of planning applications, and decisions based upon its findings and recommended policies have been upheld at national level when challenged through the planning appeal mechanism. It led directly to the extension of the central conservation area on a robust, evidence-led basis; and confirmation that the majority of the areas examined, although pleasant and contributing to the town's character, were not of sufficient special interest to merit designation. The concern for character and the policy tool of character areas have influenced subsequent policy.

However, it should not be expected that a small-scale consultancy report would have substantial and enduring impact on planning policy. In the English planning system over the past two decades, the social and environmental context has changed significantly; national and local policies, and the underlying legislation, have changed; case law and precedent have changed; and the decision-makers themselves have changed. In the quasi-judicial and adversarial planning system, only the most recent documentation carries significant weight in decision-making. The 2005 SPG has been superseded. Nevertheless it has a legacy beyond its shelf-life: it is still mentioned in decision-making, and accepted as a ‘material consideration’ even if of less significance than the current generation of policy documentation.

SoADC was unusual at the turn of the century in its interest and investment in urban morphological advice, resulting in the District Design Guidance (SoADC 2001) and the 2005 report and its supplementary planning guidance (SoADC 2005c). Yet urban morphological ideas have underpinned both the theory and practice of urban design and conservation for several decades prior to this, with urban morphology notably being ‘one of the things an urban designer ought to know’ (Moudon 1992). Stratford went for academic expertise, rather than relying on home-grown ideas of in-house staff; consultants were popular in the 1990s (Mitchell 1995). This could relate to the ‘fragmentary planning’ depending on private-sector provision of knowledge and skills that LPAs do not possess or cannot afford (Parker et al. 2018). This expenditure was reduced dramatically during the economic downturn (Waheed Nazir, Director of Inclusive Growth, Birmingham City Council, pers. comm.), but LPAs remain in a position of critical dependency with private sector consultants (Wargent et al. 2020).

Our experience with this work has also spurred further exploration of how character areas are drawn: whether there is any distinction between boundaries drawn by inspection and by measurement of characteristics such as plot dimensions (Larkham and Morton 2011). This emphasised that a morphologically trained ‘by eye’ inspection can identify boundaries as clearly as by detailed plot measurement; but the Stratford experience demonstrates the potential divergence between such morphologically informed boundaries and those delineated by political or other reasoning.

A major finding from this work, both the original study and this assessment 15 years later, is that even ‘ordinary’ areas—in this case principally interwar suburbs—have definable character. That character can be identified through detailed morphological study, which can readily subdivide streets into character areas (analogous to morphological regions: Oliveira and Yaygin 2020). This resonates with the long-standing academic interest in ‘ordinary’ areas (Groth and Bressi 1997; Jackson 1984) and that there is immense variety even within common building types such as the semi-detached house (see the Alcester and Loxley Road areas; and Jensen 2007). The same analysis can underpin decisions on whether the character is sufficiently ‘special’ to merit additional consideration such as conservation area designation. Morphological character can inform planning and management policy and can be worded sufficiently robustly to withstand challenge at local and national levels in an adversarial planning system. Even when superseded by more recent documentation, unless area character changes massively such studies will have a long-term impact. Even on this small scale, therefore, the interaction between academic urban morphology and planning policy can generate public and professional acceptance and benefit.

References

Abercrombie P, Abercrombie L (1923) Stratford-upon-Avon: report on future development. Liverpool University Press, Liverpool

Atkinson D, Cooke S, Spooner D (2002) Tales from the riverbank: place-marketing and maritime heritages. Int J Her Stud 8: 25–40

Allmendinger P, Haughton G (2019) Opening up planning? Planning reform in an era of 'open government'. Plan Pract Res 34: 438–453

Ashworth GJ, Kavaratzis M (2011) Why brand the future with the past? The roles of heritage in the construction and promotion of place brand reputations. In: Go FM, Govers R (eds) International place branding yearbook 2011. Palgrave Macmillan, London, pp 25–38

Birkhamshaw AJ, Whitehand JWR (2012) Conzenian urban morphology and the character areas of planners and residents. Urban Des Inter 17:4–17

Clifford B, Ferm J, Livingstone N, Canelas P (2019) Understanding the impacts of deregulation in planning. Palgrave Pivot, Cham

Conzen MRG (1962) The plan analysis of an English city centre. In: Norborg K (Ed) Proceedings of the I.G.U. symposium in urban geography, Lund, 1960. University of Lund, Lund, pp 383–414

Daniels R, Jones AN, Larkham PJ (1993) The character of conservation areas. Royal Town Planning Institute, London

DCLG (2012 and subsequent revisions) National planning policy guidance. DCLG, London

DCLG (2015) Explanatory memorandum to the Town and Country Planning (General Permitted Development) (England) Order 2015. http://www.legislation.gov.uk/uksi/2015/596/pdfs/uksiem_20150596_en.pdf

Department for Communities and Local Government (DCLG) (2010) Guidance on the permeable surfacing of front gardens. DCLG, London

Glendinning M (2013) The conservation movement. Routledge, Abingdon

Graham B, Ashworth GJ, Tunbridge JE (2000) A geography of heritage. Edward Arnold, London

Groth P, Bressi T (eds) (1997) Understanding ordinary landscapes. Yale, New Haven

Gunn S, Vigar G (2012) Reform processes and discretionary acting space in English planning practice, 1997–2010. Town Plan Rev 83:533–552

Heath L (2020) Jenrick announces major overhaul to planning system. Inside Housing, 12 March. https://www.insidehousing.co.uk/news/news/jenrick-announces-major-overhaul-to-planning-system-65410. Accessed 15 July 2020

Hobson E (2004) Conservation and planning: changing values in policy and practice. Spon, London

Hoyt H (1933) One hundred years of land values in Chicago. University of Chicago Press, Chicago

Hubbard P, Lilley K (2000) Selling the past: heritage-tourism and place identity in Stratford-upon-Avon. Geography 85:221–232

Jackson JB (1984) Discovering the vernacular landscape. Yale, New Haven

Jensen F (2007) The English semi-detached house. Ovolo, Huntingdon

Jokilehto J (1999) A history of architectural conservation. Butterworth-Heinemann, Oxford

Khirfan L (2010) Traces on the palimpsest: heritage and the urban forms of Athens and Alexandria. Cities 27:315–325

Larkham PJ (1990) Conservation and the management of historical townscapes. In: Slater TR (ed) The built form of Western cities. Leicester University Press, Leicester, pp 349–369

Larkham PJ (1995) Constraints of urban history and form upon redevelopment. Geography 80:111–124

Larkham PJ (1996) Conservation and the city. Routledge, London

Larkham PJ (1999a) Conservation and management in UK suburbs. In: Harris R, Larkham PJ (eds) Changing suburbs: foundation, form and function. Spon, London

Larkham PJ (1999b) Tensions in managing the suburbs: conservation versus change. Area 31:359–371

Larkham PJ (2004) Conserving the suburb: mechanisms, tensions, and results. In: Stanilov K, Scheer BC (eds) Suburban form: an international perspective. Routledge, New York, pp 241–262

Larkham PJ, Chapman D, Morton D, Birkhamshaw A (2005) Stratford upon Avon: examination of residential character. Unpublished draft report. SoADC, Stratford upon Avon. https://www.stratford.gov.uk/doc/208345/name/Introduction%20and%20Map%20unlocked.pdf/. Accessed 15 July 2020
Larkham PJ, Morton N (2011) Drawing lines on maps: morphological regions and planning practices. Urban Morphol 15:133–162
Martin GH (1968) The town as palimpsest. In: Dyos HJ (ed) The study of urban history. Edward Arnold, London, pp 155–170
Massey D (1995) Places and their pasts. Hist Workshop J 39:182–192
Miller C, Sandford E (2020) Most expensive streets in Coventry and Warwickshire in 2019 revealed. Coventry Telegraph, 3 March
Mitchell VW (1995) A survey of chief planning officer attitudes towards planning consultants. Int J Public Sector Manag 8:20–34
Moudon AV (1992) A catholic approach to organising what urban designers should know. J Plan Lit 6:331–349
Oliveira V, Yaygin MA (2020) The concept of the morphological region: developments and prospects. Urban Morphol 24:35–53
Parker G, Street E, Wargent M (2018) The rise of the private sector in fragmentary planning in England. Plan Theory Pract 19:734–750
Pendlebury J (2008) Conservation in the age of consensus. Routledge, London
Pendlebury J, Strange T (2011) Urban conservation and the shaping of the English city. Town Plan Rev 82:361–392
Sabelberg E (1983) The persistence of palazzi and intra-urban structures in Tuscany and Sicily. J Hist Geogr 9:247–264
Skea R (1996) The strengths and weaknesses of conservation areas. J Urban Des 1:215–228
Smith L (2006) The uses of heritage. Routledge, London
Stratford on Avon District Council, SoADC (2001) District design guide. SoADC, Stratford upon Avon
Stratford on Avon District Council, SoADC (2002) Stratford Town Design Statement. SoADC, Stratford upon Avon
Stratford on Avon District Council, SoADC (2005a) Residential design guidance and review of conservation areas: report to the Executive of the District Council, 4 July 2005. SoADC, Stratford upon Avon
Stratford on Avon District Council, SoADC (2005b) Avon Area Community Committee Meeting minutes, 24 March 2005. http://democratic.warwickshire.gov.uk/cmis5/PreviousCommittees
Stratford on Avon District Council, SoADC (2005c) Design in residential areas supplementary planning guidance. SoADC, Stratford upon Avon
Stratford on Avon District Council, SoADC (2006) Stratford-on-Avon District local plan review 1996–2011. SoADC, Stratford upon Avon
Stratford on Avon District Council, SoADC (2018a) Stratford-upon-Avon area transport strategy. https://www.stratford.gov.uk/doc/206646/name/Adopted%20Stratford%20Area%20Transport%20Strategy.pdf. Accessed 15 July 2020
Stratford on Avon District Council, SoADC (2018b) Stratford-upon-Avon neighbourhood plan. https://www.stratford.gov.uk/planning-building/stratford-upon-avon-neighbourhood-plan.cfm. Accessed 15 July 2020
Tait M, Inch A (2016) Putting localism in place: conservative images of the good community and the contradictions of planning reform in England. Plan Pract Res 31:174–194
UK Government (2004) The Town and Country Planning (Local Development) (England) Regulations 2004. http://www.legislation.gov.uk/uksi/2004/2204/contents/made. Accessed 15 July 2020
Uzzell DL (1996) Creating place identity through heritage interpretation. Int J Herit Stud 1:219–228
Vâlceanu DG, Kosa RA, Tamîrjan DG (2014) Urban landscape as palimpsest. Urbanism Arhitectura Constructii 5:17–26

Vines A (2018) The wringing of hands. Context 154:42–43
Wargent M, Parker G, Street E (2020) Public-private entanglements: consultant use by local planning authorities in England. Eur Plan Stud 28:192–210
While A (2007) The state and the controversial demands of cultural built heritage: the origins and evolution of post-war listing in England. Environ Plan B 34:645–663
While A, Short M (2010) Place narratives and heritage management: the modernist legacy in Manchester. Area 43:4–13
Whitehand JWR, Carr CMH (2001) Twentieth-century suburbs: a morphological approach. Routledge, London
Whitehand JWR, Larkham PJ (1991) Suburban cramming and development control. J Prop Res 8:147–159
Whitehand JWR, Larkham PJ, Jones AN (1992) The changing suburban landscape in post-war England. In: Whitehand JWR, Larkham PJ (eds) Urban landscapes: international perspectives. Routledge, London, pp 227–265
Wright P (1985) On living in an old country. Verso, London
www.gov.uk/government/collections/planning-application-statistics

Part II
Urban Morphology and Urban Design

Chapter 4
Towards an Eclectic Urban Morphology

Ivor Samuels

Abstract This chapter will examine two planning instruments. Both were adopted by their commissioning planning authorities and their preparation incorporated approaches from the urban morphology spectrum. The *Plan d' Occupation des Sols* (POS) for Asnières sur Oise was adopted by the Commune of Asnières sur Oise in 1992. The Oxford West End Design Code (OWEDC) was adopted by the City of Oxford in 2008 as part of its West End Action Area Plan 2007–2016. A brief résumé of the POS is followed by a discussion of the reasons for its abandonment as revealed by a research study. The lessons from this case were subsequently incorporated in the OWEDC. In spite of some of its basic assumptions being rendered obsolete by the economic crisis of 2008, the principles embodied in OWEDC have continued to inform subsequent planning decisions for the area. In conclusion, the obstacles to incorporating urban morphological approaches in planning practice, including the need to incorporate considerations of economic feasibility both in the preparation and the implementation of the proposals, are discussed as are the benefits of linking academic work with practice.

Keywords Design codes · French and UK planning · Different morphological techniques

4.1 Introduction

Following a brief review of the significance of 'the English Townscape Group' (Rapoport 1977) with respect to matters of urban form in British urban design practice in the last quarter of the twentieth century, this chapter will examine two planning instruments which attempted to extend this approach. Their preparation by British-based practitioners was unusual in that it incorporated concepts from the urban morphology spectrum in addition to Townscape. The *Plan d'Occupation des Sols* (POS) for Asnières sur Oise was adopted by the French Commune of Asnières

I. Samuels (✉)
University of Birmingham, Birmingham, UK
e-mail: ivor.samuels@googlemail.com

V. Oliveira (ed.), *Morphological Research in Planning, Urban Design and Architecture*, The Urban Book Series, https://doi.org/10.1007/978-3-030-66460-2_4

sur Oise in 1992 and the Oxford West End Design Code (Placemaking Associates et al. 2007) was adopted by the City of Oxford in 2008 as part of its West End Action Area Plan 2007–2016. Barke (2015) has noted how political and economic contexts distinguish practice from research and it is therefore considered important to discuss these influences since both instruments were prepared in different legislative and political contexts and under different professional circumstances.

In architecture and town planning teaching and practice in the United Kingdom was dominated by the work of the Townscape School, mainly through the work of Gordon Cullen (1961). His approach celebrated the virtues of historic towns, notably through the easily understood technique of serial vision, and contrasted these with the environments which had been produced by the programme of post-war reconstruction. This had been savagely criticised in the Architectural Review under the editorship of Ian Nairn who coined the term subtopia and published a special issue entitled Outrage (1956), illustrated by Cullen. Revisiting this influential book, it is significant to note that, while there are scores of photographs and drawings on every page, there are no plans of the locations they illustrate. The other contemporary fundamental reference was the work of Kevin Lynch (1960).

In their map of contributions to urban form, Gautier and Gilliland (2006) include Lynch on the Externalist and Normative axes but do not include Cullen, who would presumably be located along the same axes. Both these authors offered practising professionals a kit of techniques which could be easily understood, adopted and used as design tools by architects and urban designers and they had a considerable influence on urban design practice (Larice and MacDonald 2007). They also had the great advantage of being easily understood by laypersons, which facilitated community participation.

In the fields of architecture and urban design in Britain the period was notable for its neglect of those geographers concerned with urban form such as Smailes or Stedman cited by Larkham (2006) in his account of urban morphology in Britain. As a Masters student on an Urban Design and Regional Planning course in the Faculty of Social Sciences (the base for the study of geography) at Edinburgh University in the late 1960s, this author only received one or two lectures from a geographer on the work of Burgess and Park.

The concepts of Cullen and other members of the Townscape Group were fundamental for the implementation of the Civic Amenities Act of 1967. This was based on French legislation for the *Secteurs Sauvegardés* and was the first time in Britain that parts of urban areas or entire settlements, as opposed to single buildings, were identified to receive special consideration in the planning system. In 1968 major studies were produced for Bath, Chester, Chichester and York by the Ministry of Housing and Local Government as models for the type of instrument which could be produced for the newly designated conservation areas, and in 1969 Worskett produced a more general work demonstrating a possible approach.

There is no indication that the work of Conzen (his fundamental work on Alnwick had been published in 1960) or that of any of the other geographers noted by Larkham, was known to the authors (mainly architects) of these works. This omission demonstrates the rigid boundaries within which professions worked at that time. This can be partly explained by the numerous specialised professional divisions that exist in Britain when compared with other contexts, with town planners having established an independent identity early, relative to their status in other countries, in addition to the various specialisations within the Royal Institute of Chartered Surveyors. Urban design had to be invented as a discipline (but not as a profession) and postgraduate degrees were set up in the 1970s as a result of a growing concern for the professional division between architects and town planners. The latter in their work are more likely to be involved with urban form than architects, even though planning courses according to a survey by the Town and Country Planning Association in 2013 relegated to minor position topics of design in favour of legal and procedural matters. Another reason for the neglect of the urban environment by architects may be the obsession of architectural education with special buildings as opposed to the ordinary buildings which make up most of our towns. While the International Seminar on Urban Form (ISUF) has been successful in bridging between professions internationally—e.g. British geographers with Italian architects, perhaps it has been less successful in bridging across disciplines within each of its national boundaries.

This author, while working as an architect in Milan, became aware of other approaches to urban form being formulated by Italian architects who were systematically investigating urban form in a way which was very different from the Townscape School's neglect of underlying structures such as plot systems or street networks. These included Muratori in Venice and Cervellati in Bologna and such eminent practitioners as Rogers and Rossi. They were part of a general questioning of the tenets of modernism and the rediscovery of the virtues of inherited urban forms. It was at the Joint Centre for Urban Design (JCUD) at Oxford Polytechnic, working with Latin American postgraduates whose linguistic skills also enabled them to refer to Italian sources, that a geographer colleague pointed out a similar approach was being led by Jeremy Whitehand in the School of Geography at the University of Birmingham. Contact was made and he organised a seminar in Birmingham in 1982 which for the first time brought together the work of the Italian and British schools of urban morphology. Around the same time Gianfranco Caniggia made a great impact when he spent a month in Oxford in the same year that he went to Seattle, invited by Anne Vernez Moudon.

In the UK, outside the field of conservation, Townscape proved a very effective instrument and offered a widely adopted set of tools for the design guidance of new housing, which had been its original motivation for Nairn (1956) and Cullen (1961). Of these guides perhaps the most widely emulated was the Essex Design Guide of 1973 (updated in 2015 and 2018) which pioneered attempts to retain the character of localities in speculative housing (ECC 1973). It was this experience which was to be of great relevance to the first of the case studies discussed. Drawn principally from architecture and town planning backgrounds, urban designers are firmly based on the normative side of the Gautier and Gilliland spectrum. They could be considered as mediators between the triumvirate of landowner, funder/developer and regulator/planner whether these are in the public or private sector. Although the relative influence of each member of this triumvirate varies according to the political/economic context, their power over the most significant choices to be made in urban development, whether directly through implementing building projects or by the formulation of instruments (codes, plans, etc.), is much greater than that of the urban designer or the architect (McGlynn 1993). In order to devise effective projects or instruments, professionals have therefore to understand the political/economic context as well as the urban form context in which they are working.

Practice is orientated towards action rather than academic reflection but practitioners draw on the insights and methods of academic researchers in addition to those of other professionals. They have to be sufficiently familiar with the very wide range of approaches to understanding urban form and its transformation so as to be able to select for each project those which they consider most appropriate for the task at hand. Therefore, it is important to be familiar with the way different approaches have been applied in practice and their effectiveness. Unfortunately, evaluations of this type are relatively rare and it is unusual to have access to the sort of evaluation which took place of the first of the projects below.

While Townscape is useful because of its accessibility, particularly in discussions with lay stakeholders, it does not penetrate behind the facades of buildings to explain the structure which determines these facades. Recourse to other methods is necessary. At Asnières a hybrid of Cannigia and Conzenian approaches set out a framework for control through plot types. However, in the same project an attempt to control building detail failed. Probably such a high degree of control can only be effective where there is a continuity of landownership as happens at Poundbury, the Prince of Wales' urban extension to Dorchester (Samuels 2020).

Another example of the application of different methods was the adoption of the Space Syntax concept of cones of vision in the West End design code to indicate locations where buildings and /or floorscape should be varied. However, a Space Syntax analysis of connectivity undertaken for this project produced proposals for shifting the location of an existing street which was politically and economically untenable and therefore excluded from the code as adopted. In the same project the Regulating Plan was taken from New Urbanist practice even though the broader deficiencies of this approach with respect to urban morphology have been noted (Scheer 2008).

4.2 The *Plan D' Occupation Des Sols* (POS) for Asnières Sur Oise

The commissioning of this work from a team of staff and students of the School of Architecture at Oxford Brookes University (OBU), had its origins in a number of projects carried out over a decade from 1981 following an initial contact made by Paul Oliver with the *Patrimoine Historique et Artistique de la France*, whose president was the late Mme Paulette Abravanel. This NGO sponsored studies in small communes (there were 37,000 communes in France at that time), where teams of university students and staff were accommodated by the local inhabitants for up to three weeks of fieldwork. These projects or *Actions Pilotes* ranged from the entirety of small villages e.g. Donzenac (Samuels et al. 1987) to parts of larger towns e.g. Provins 1981 (Joint Centre for Urban Design 1980)

The staff/student teams varied in size and they worked on different aspects of the project. The work of the students was assessed as part of the academic curriculum. The resulting reports were up to 100 pages long—e.g. for the Ville Basse at Provins, a town of 11,000 inhabitants where the upper town had statutory protection, but the heritage of the lower town had been ignored. Other reports for smaller communes were much shorter, but resulted in projects being implemented after the teams had left—e.g. the market hall built at Donzenac with only 2,000 inhabitants.

Asnières sur Oise, with a population of 2,400 people, is a commune of 1,400 hectares, 35 km north of Paris, in the Val d'Oise with an attractive location on the northern escarpment of the Plaine de France. The Action Pilote, presented and exhibited to the citizens, the public, the Town Council and the Mayor at the end of the OBU project, included embellishment of the public spaces, enhancement of the settlement edges, outline proposals for an abandoned industrial area and a framework for a design guide. It was this last part of the project which was to be developed into a *Plan d'Occupation des Sols*, although at this stage of the work, it would have carried little conviction without the other supporting studies (Samuels 1993).

The Mayor, Paul Lassus, by profession an international commercial lawyer, was very concerned that his Commune, located on the edge of the Paris conurbation, would become just another *banlieu* or suburb and the characteristic new development of *pavillons* (small detached houses) would completely erase its distinctive character as had occurred with other small communes in similar locations. In particular, he was concerned with the manner of expansion of the neighbouring commune of Viarmes right up to his commune boundary with extensive areas of detached housing (see Google Earth, 49º07′37.16″N, 2º22′11.93″E).

At that time development was controlled by the POS. This was a document which was legally binding and if a proposed development was in accordance with its regulations, then it had to be approved. There was no discretion to take other factors into consideration as in the British planning system. It was therefore the most significant instrument for determining the form of future development. However, these regulations, as expressed in the POS, were normally limited to establishing future land use zones and the quantity of development allowed within these. This was set out as a maximum *Coefficient d'Occupation des Sols* (COS) or Floor Area Ratio (Merlin and Choay 1988). There was no consideration of the quality of the development proposed.

This problem had been realised and there was an emerging attempt to promote a wider view of the purpose of the POS, although at that time there were few models of how this might be achieved. This can be explained by the way plans were usually prepared. The communes were usually too small to employ permanent in house technical staff and, although they were able to commission the work from independent consultants, financial considerations meant that they usually took advantage of the technical assistance offered by the local offices of the *Direction Départemental de l' Equipment* (DDE), the Ministry in charge of planning matters. However, these were mainly staffed by engineers and administrators, who were more concerned with administrative efficiency than with local character. The 1987 POS for Asnières, prepared by this agency, was of a standard form which neglected the particular characteristics of the village.

The Mayor had been newly elected on a platform of improving the quality of the environment of the Commune and he was conscious that the POS he inherited would not allow him to achieve his ambitions. He was an admirer of the interventions of the Prince of Wales at that time through such initiatives as Poundbury and his published work (HRH Prince of Wales 1989) and he realised that the Action Pilote offered an opportunity to achieve a different type of POS, which would be concerned with quality and not only quantity. Following the presentation of the students' work he therefore dismissed the architect who had been commissioned to prepare the plan and invited this author to put together a team to prepare a qualitative POS for his commune.

The Mayor was able to prepare a new plan under a decentralisation law of 1983 which allowed even the smallest communes to prepare their own POS, although only two communes in the Department of Val d' Oise had taken up that possibility. There were two agencies of central government which could have supported him but neither of them would do so. In certain Departments the *Conseil d'Architecture, d'Urbanisme et de l' Environnement* (CAUE) had prepared design guides for traditional architecture, however in this case the degree of control that was being proposed, was considered to be undemocratic and too restrictive of architects' creativity—a familiar argument to the proponents of design guidance. The *Architectes des Bâtiments de France* (ABF) took a similar view. Because there were two protected buildings in Asnières, the ABF had to be consulted about any development within a 500-metre radius of these. This included most of the built up area of Asnières.

The 1987 POS followed the established four-part formula, starting with a *Rapport de Presentation* (survey), setting out the physical, demographic and economic context leading to a set of general development policies. The second part was a 1:10,000 zoning plan for the whole commune with 1:2,000 scale inserts for the built up areas followed by the *Reglements* (regulations) for the zones defined on the plans. The final part was an Appendix which covered such matters as future road improvements and sites reserved for public infrastructure investments.

The survey section of the earlier plan was very general and divided the commune into two zones, urban and natural. Within the former there were four sub-zones defined according to a mixture of morphological and land use criteria. Thus, the older area of the village was incorporated into one zone of continuous building for housing, services, trade and industry. All the newer areas were allocated for individual houses with a narrower range of uses. Larger individual houses were zoned separately. The fourth zone covered two industrial sites. Within all zones permitted and excluded land uses, and permitted plot ratios were defined. The regulations included siting in relation to roads, parking provision, building heights and external appearance, in fact all those aspects an urban designer would consider necessary to control urban form.

The first problem with this POS was that it did not set out any rationale for the selection or rejection of different detailed provisions, which were considered important in order to maintain the essential characteristics of Asnières. This neglect had made the Mayor's attempts to influence negotiations with developers seem to be subjective and arbitrary. The second problem was the lumping together of all the older areas of the village, which reduced the subtle but important differences in form and materials between the different tissues, so that the use of the same regulations for all the older areas reduced their diversity. Where an attempt had been made to formulate specific regulations such as those dealing with appearance, these often were in such vague terms as 'should be coherent' or 'in harmony with' the traditional building forms.

The new 1992 POS had to be an instrument which could be used to demonstrate to laypersons, professional builders and architects the characteristics of the village and then derive from these a range of possible solutions applicable to their specific development. Thus it was similar in intention to the UK design guide but, unlike that type of advisory document, it had to be legally binding and capable of surviving challenges in the Court of Administrative Law. The didactic quality of the design guide was very appealing to the Mayor; he was particularly attracted to the possibility of illustrating both acceptable and unacceptable solutions to development problems. However, in respect to this quality, the team was concerned by the lack of rigour in UK Design Guides since these often consisted of subjective responses to place, which would be difficult to uphold against a legal challenge. The task was to define the characteristic qualities of the settlement in rigorous terms of built form and then formulate equally careful prescriptions which would ensure the future continuity of these qualities. To achieve this, it was necessary to work within the context, unfamiliar to the Oxford team, of the *Code d' Urbanisme* and the Civil Law in general, where it deals with property rights, and also within the constraints of political feasibility, so as to ensure the Mayor's future re-election. The only way in which these requirements could be met was to incorporate the Mayor with his legal skills as an integral part of the team. While in some contexts this might appear to be an irregular procedure, in France it is not unusual because there is much less distinction between political and technical roles.

Similar to its 1987 predecessor the new POS was also in four sections, but the first three were radically different from those of its predecessor. It was necessary to adopt a methodology that enabled a systematic argument to be built up, which led from the analysis of the qualities of Asnières in the first part (*Rapport de Presentation*) via the plans in the second part to the regulations in the third part. As opposed to the quantative POS, dominated by considerations of the amounts and types of activities, which were permitted, it was to be a qualitative POS primarily concerned with quality of place.

The method adopted was initiated by Karl Kropf who was a member of the team, even though he was no longer a Masters student at the JCUD, but working on his PhD at the University of Birmingham (Kropf 1993). He used Asnières as a case study in his dissertation and devised a simplified method based on levels of resolution derived from the work of Conzen and Cannigia. Thus, the way the streets, blocks, plots, building forms and construction elements, tree types and façade colours determined the character of the different quarters of the settlement, was systematically described in the first part of the POS (*Rapport de Presentation*). These studies enabled seven distinct quarters to be identified which approximated to groupings of plan units or tissues. Each quarter was illustrated by examples of existing development which demonstrated the way they contributed either positively to, or detracted from, the distinctive character of each quarter (Fig. 4.1). These examples ranged from the position of buildings on plots to showing, at the Mayor's insistence, acceptable and unacceptable eave, window shutter and fencing details (Fig. 4.2).

The analysis was undertaken by a combination of direct observation, discussion with local experts and desk research into the local vernacular. The team had the benefit of access to a comprehensive set of cadastral maps held by the commune. These included those dating from the eighteenth and nineteenth centuries. The survey concluded with short sections on demographics and employment noting that only a small number of people worked in the commune, which suggested a priority for the reuse of an abandoned and derelict light industrial area in the centre of the village.

In the new regulations the range and quantity of uses prescribed approximated very closely to those in the 1987 POS with the exception of the industrial site where an area of residential floor space was required in order to foster a mix of use. Development was to be restricted to sites alongside roads indicated on the zoning plan. This measure was introduced by the Mayor to prevent the subdivision of larger sites by the use of *cul-de-sacs*, because the Civil Code permits the subdivision of plots every ten years. It was considered necessary in order to prevent speculative developers from acquiring large sites and then dividing them into small plots for detached houses—the *lottissements*.

Each of the seven urban zones in the built up areas (zoned as Urban Areas, UA), was named according to their character and had a separate section devoted to it of around ten pages in the final document. These urban zones were The Village (UAa), The Farms (UAb), The Hamlets (UAc), Baillon (the name of a hamlet detached from the main settlement (UAd), The New Quarters (UG), Delacost (the industrial area UI), and a zone of Older Properties and their Parkland (UNH). One section covered the construction elements of all the buildings in all of the UA zones.

For each zone a range of acceptable plot types was set out with minimum dimensions, plot proportions, buildable area and plot coverage. The permitted building type varied according to the position of the plot in the block-front, side or corner (Fig. 4.3). The resulting framework within which a developer had to work was unusual in that it offered a relatively wide choice at the more structural level with several possibilities for plot sizes and the arrangement of buildings on them, but a limited range of choice with respect to the construction details. This mirrors vernacular buildings which historically were restricted to the limited range of materials available locally. This is in contrast to the normal procedure of contemporary developers where, for economies of scale, the same building type is enveloped with a variety of materials of different types or colours, not necessarily derived from local practice, in an attempt to demonstrate either a response to local character or to achieve a degree of variety (McGlynn and Samuels 2000).

19

3- Comment le parcellaire commande la forme des constructions

La parcelle est l'unité de base de division du foncier ainsi que l'unité de base des îlots. Une agglomération grandit parcelle par parcelle. Une parcelle doit se considérer d'un point de vue tri-dimensionnel: c'est une surface sur laquelle s'élèvent des constructions et des murs de clôture. Ce sont les constructions et les murs de clôture qui définissent physiquement la parcelle et la divisent en espaces utiles. La façon dont ces constructions et ces murs de clôture sont organisés contribue à la forme et au caractère de l'îlot et donc du quartier et de l'agglomération toute entière.

Rue Pierre Brossolette: rue de parcelles de type traditionnel

A Asnières et Baillon il existe plusieurs types de parcelle identifiables. Le type le plus répandu est sans doute la parcelle à façade étroite sur laquelle on a édifié une maison parallèlement à la rue en bordure d'alignement. Ce type, quand il se répète en série le long des rues, crée l'espace familier de la rue de village, avec des portes d'entrée de maisons d'habitation et des devantures de boutiques ouvertes sur la rue.

. *Rue de parcelles de type pavillonnaire*

La parcelle de type pavillonnaire se situe en opposition par rapport au type qui vient d'être décrit. Elle est relativement petite, généralement carrée et supporte une habitation monofamiliale d'un ou deux étages implantée en son milieu. Là où ce modèle est généralisé, l'îlot qui en résulte ne crée pas de séparation entre l'espace privé et l'espace public. De plus le modèle pavillonnaire est plus consommateur d'espace que le modèle du village traditionnel.

Fig. 4.1 POS, survey report showing traditional street in centre and new pavilions (*Source* Mairie d'Asnieres sur Oise et al. 1992)

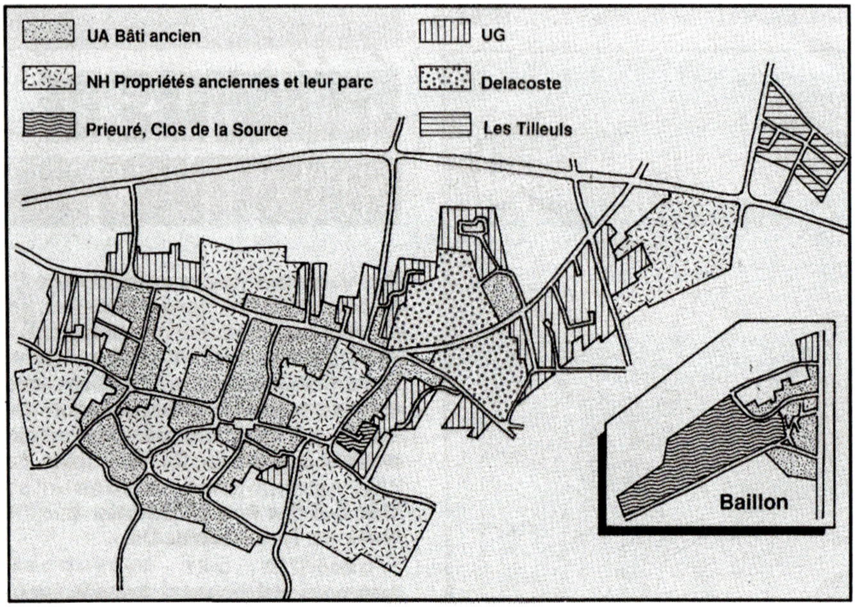

Fig. 4.2 POS, plan of general zones of Asnières (*Source* Mairie d' Asnières sur Oise et al. 1992)

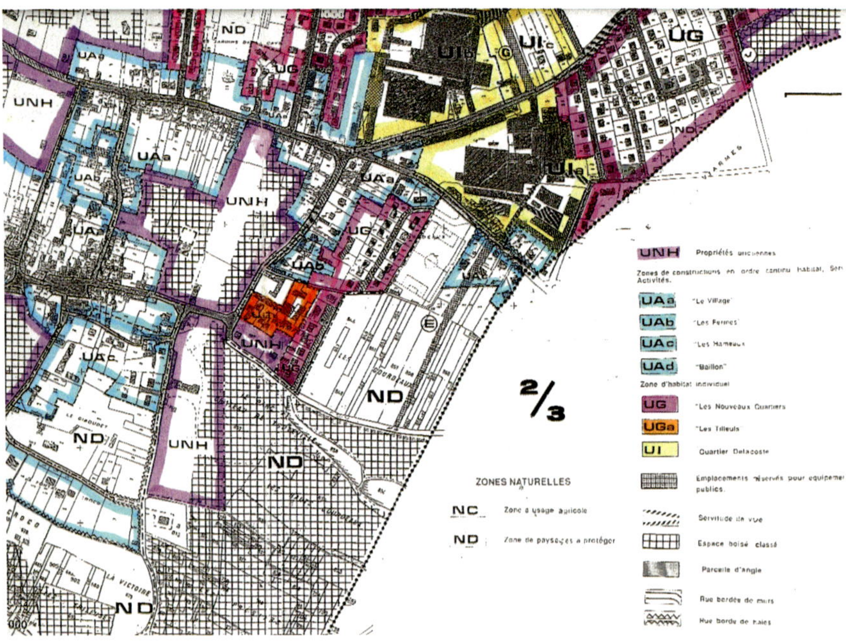

Fig. 4.3 POS, plan of the southeast section of Asnières showing typological zoning (*Source* Mairie d' Asnières sur Oise et al. 1992)

4.3 The Application of the Asnières POS

In 1995, Laurence Pattacini, a landscape architect qualified in France, undertook a study entitled 'A better place to live' in partial fulfilment of her MA in Urban Design at the JCUD (Pattacini 1996). She examined several cases of POS and *Zones d' Aménagement Concerté* (ZAC) and used the contacts established four years earlier to follow up in detail the impact of the 1992 POS of Asnières.

Her interviews with different actors in the process of creating and adopting the POS confirmed that the arrangements for presenting the work in progress regularly to the inhabitants and elected members, supported by the circulation of easily understood documents (Fig. 4.4), ensured a process which was both democratic and didactic and which changed their perception and understanding of their living environment (Pattacini 1996, p. 38). This was confirmed by the continuation of the broad principles of the plan even after a new Mayor was elected in 1994 in opposition to Paul Lassus. While there was reduction in the amount of building because the popular *lottissement* type of development was no longer possible, two buildings demonstrated the application of the new rules. These were a school built by the commune and a private house.

During the implementation of the POS the main problems and conflicts with the inhabitants related to their opposition to the building details, choice of materials and colours. An architect working with the Historic Monuments considered that, because of the complication and expense resulting from the imposition of the prescribed details—e.g. for cornices, a degree of economic deprivation was imposed on the inhabitants (Fig. 4.5). While under the mayoralty of Paul Lassus the POS was administered through the employment of an architect, Stefan Ferrazzini, who was sympathetic to its principles, under his successor the administration was taken over by an officer of the DDE who was not based in Asnières. She is reported as objecting to the complication of the regulations and the time it took her to process planning permissions. This reveals the difficulties administrators have when confronted with a nonstandard document. The Mayor's concern for the future of his commune is justified by the nature and extent of development since his POS was prepared. Recent Google Earth images reveal a sprawl of development beyond the boundaries of the old town consisting almost entirely of detached houses—the *pavillons* so abhorred by Paul Lassus.

On a more positive note, in 1995 the Asnières POS was awarded the *Prix National de l' Innovation* by the *Caisse Nationale de Crédit Agricole* and the *Conseil Supérieur du Notariat*. This may have been partly in acknowledgement and celebration of the way a lawyer had been so central to its production, which marked a return of that profession to central role in town planning. In the middle ages notaries had been responsible for the layout of new settlements such as the English and French *bastides*—e.g. Pons Maynard who laid out Montreal-du-Gers (Reps 2016).

As a result of the work on this POS the Mayor, Paul Lassus, set up a consultancy firm 'Qualipos' in his legal office in Paris and it prepared a number of plans for communes in different parts of France based on the Asnières project. The office was

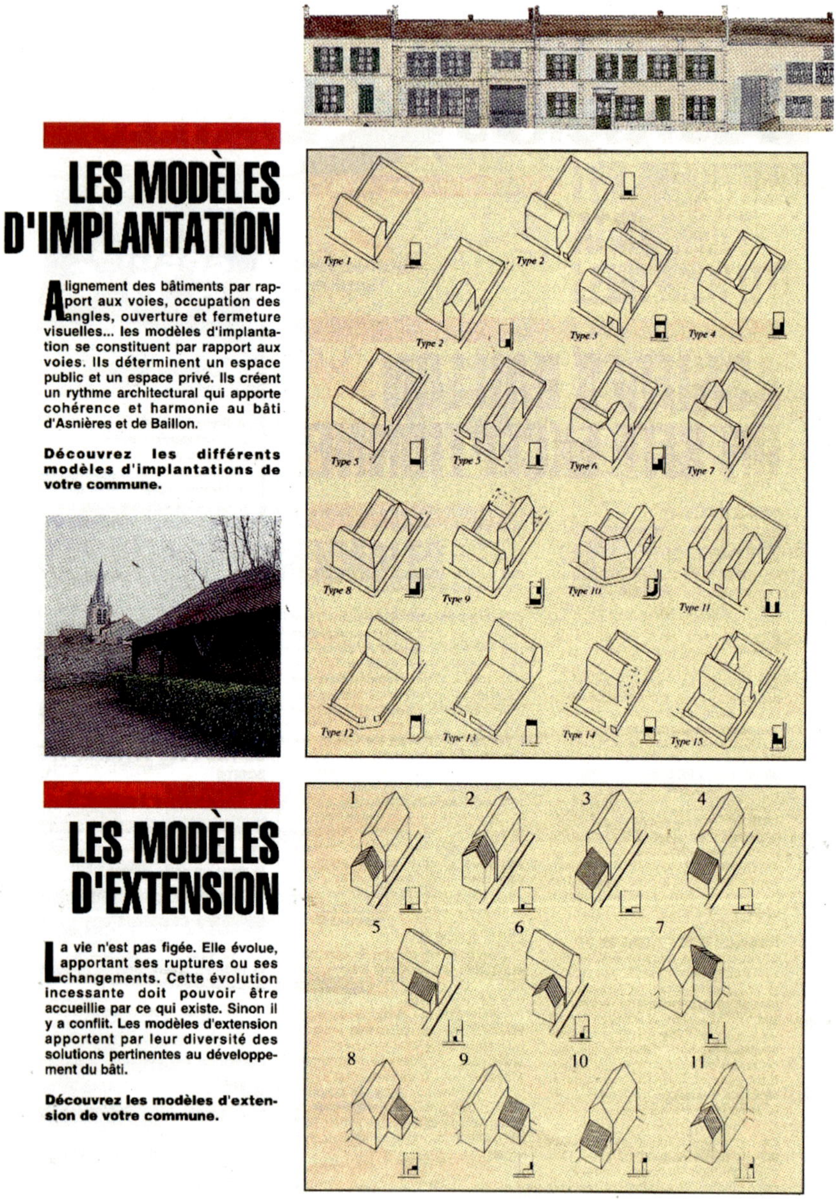

Fig. 4.4 POS, leaflet circulated to inhabitants of Asnières showing a range of possible building types and their locations on plots and (below) possible forms of building extension (*Source* Mairie d' Asnières sur Oise 1992)

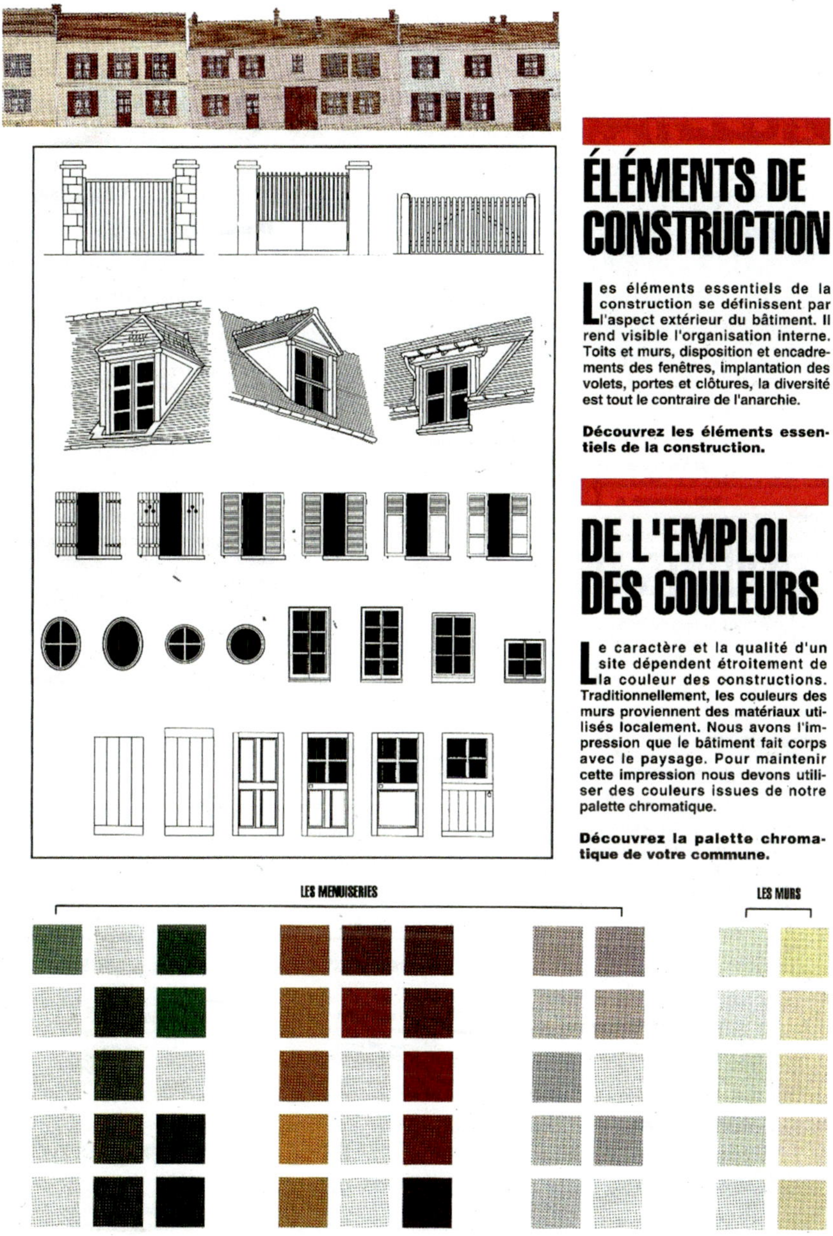

Fig. 4.5 POS, leaflet circulated to inhabitants of Asnières showing permitted designs for gates, windows and doors and the range of colours permitted (*Source* Mairie d'Asnieres sur Oise 1992)

staffed by young architects employed on a minimum salary as established by the French Government. The authors of the Asnières POS had a *pro bono* advisory role in these projects. In addition to the commune of Montreal-du-Gers, mentioned above, these included Mennecy, Boigneville, Mesnil-Saint Père, Aumont en Halatte and St. Gervais (Samuels 1999). This last POS was for a settlement on the lower slopes of Mont Blanc; it extended the work at Asnières because in that mountainous area the building forms related to topography and access rather than to plot subdivisions. It also involved three dimensional proposals in the form of a *plan masse* for new development (Samuels and Pattacini 1997). In spite of the survey work undertaken for all these plans being much less detailed than that at Asnières, the support of a subsidised office in Paris and relatively low labour costs, these projects proved financially unviable and Qualipos ceased trading after two years.

4.4 Oxford West End Design Code

This design code was one of the nine appendices in the second of the two-volume West End Action Area Plan (WEAAP) for an extensive area of Oxford lying adjacent to the historic city centre. It is the location for a large Further Education College in a number of buildings and a range of other facilities such as a free-standing skating rink, a large car park and small light industrial buildings. Part of the area includes parkland which borders the River Thames. The four aims set out in the WEAAP were to create an attractive network of streets and spaces, high quality built environment, a strong balanced community and a vibrant and successful West End. Clearly, the Design Code was central to the achievement of the first two of these aims.

Sue McGlynn and Ivor Samuels as Place Making Associates were invited in 2007 to prepare the Oxford West End Design Code (OWEDC); the rest of the plan was prepared by Oxford City Council. McGlynn and Samuels had been teaching colleagues at the JCUD and McGlynn had been an urban designer in the Planning Department of Oxford City Council and was one of the authors of Responsive Environments (Bentley et al. 1985). They were assisted in the latter stages of the OWEDC by several recent graduates from the JCUD.

It was envisaged that between 600 and 800 new dwellings would be built in this area, but the need for a mix of uses was also emphasised which included retailing, student accommodation, community and tourist facilities and business accommodation. The uses were not allocated to the 24 sites identified for development, but a table showed for each site the range of uses which were considered appropriate.

Typically, design codes are produced for specific programmes of development, e.g. to implement a masterplan. Where the area being coded is more extensive with no fixed plan then the instruments usually fall into the category of design guides. In the case of the West End the uses were not fixed, but there was in existence a defined boundary, a number of sites identified for development and a network of existing streets which needed to be extended. It was therefore decided to focus the coding on the street system, the most enduring of urban morphological elements.

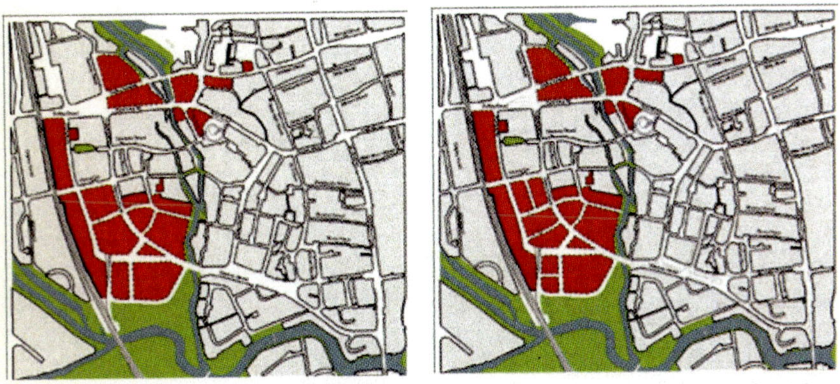

Fig. 4.6 OWEDC, Essential and optional street mesh showing different development parcels (*Source* Oxford City Council 2008)

Since uses were not decided it was not possible to define a fixed network of blocks because their size would need to vary according to the functional demands. This code therefore chose to set out an 'essential mesh' defined by the existing streets and their extensions, which would be capable of accommodating the more extensive commercial and leisure projects. In the event of the proportion of residential uses increasing relative to these activities, an 'optional mesh' was laid out, which was capable of accommodating more housing (Fig. 4.6).

The British planning system leaves many major decisions about form to the producers of housing. It is a historic tradition which resulted in such admired developments as the London squares, which were laid out by major landowners in the eighteemth century. Recognising this the authors had been anxious to engage directly with developers and had been involved in the delivery of a series of courses to the seven regional design teams of Wilcon Homes, at that time one of the major British housebuilders with a programme of around 4,500 annual house completions. The Wilcon Marketing Director had been anxious to improve the quality of his product and had been influenced by visiting the United States where he visited developments designed by New Urbanist architects (McGlynn and Samuels 2000). It was at a time when there was a growing concern to raise the quality of housing through the promotion of such British Government publications as 'By Design' (DETR 2000). Part of this concern was a quest to understand the character of the locality into which the new housing was being inserted and reinterpreting its qualities in the new project. It was clear that the sort of studies undertaken at Asnières would not be feasible within the normal resource capacity of either private consultancies or public planning offices. A simplified approach was developed and presented to the Wilcon design teams which focused on the importance of the street and its design as the main generator of urban form and this was the approach adopted for the OWEDC.

The problems of making operational the Asnières POS because of its complexity and the experience of working with housebuilders led, in the OWEDC, to an emphasis on producing a document which was easily understood and accessible to different stakeholders. It was set out in five sections and the user was guided through these in a sequence of steps corresponding to the levels of resolution from the street mesh of the area to the design of individual buildings (Fig. 4.7).

Section A (Step 1). The first step was the location of the development parcel or plot on the Regulating Plan. This, a fundamental element of US New Urbanism practice (Form-Based Codes Institute 2020), is the key source for the general provisions and standards relating to the streets adjacent to the proposed development. Both existing and proposed routes are classified according to the likely intensity of use and movement ranging from main streets (Type 1) to pedestrian cycle and riverside routes (Type 4)—Fig. 4.8. The essential street mesh is intended to ensure the connection of the West End with the City Centre with an optional mesh to be adopted if there is an increase in residential development. The distance between plot frontage lines is defined according to the street types set out in Section B and the proposed development is required to address the highest order street frontage so as to ensure continuity of building frontage along these. This section also deals with the siting of large buildings on the largest blocks as defined by the Regulating Plan. It also shows the way these structures must be set back so as to allow buildings with more active frontages to be located on the surrounding streets.

Section B (Step 2). The identification of the relevant street type, which sets out the scale of development which is appropriate as well as street width and building height, the continuity of building frontage and the degree of active frontage to the public realm. As in other sections of the OWEDC this was illustrated wherever possible by examples drawn from Oxford (Fig. 4.9). There was a concern that often design guides and codes were illustrated by examples from other places which could be claimed as irrelevant to the local context. By showing examples with which they were familiar, they allowed local stakeholders to understand easily the underlining rationale of the code.

In order to create safe, lively and interactive streets it is necessary to ensure that they are lined by active frontages. To achieve this, in addition to defining degrees of continuity of frontages and range of permissible set back distances of building lines, the number of required doors per 100 metres (adapted from Gehl et al. 2006) was set out for various street types together with the percentage of transparency, in terms of window openings measured across ground floors (Fig. 4.10).

Section C (Step 3). Locating the development in relation to the street segment which is the main device in the code for achieving a degree of variation for the general provisions. A criticism of many design codes is that they result in an unacceptable degree of uniformity. Alternatively, they result in an apparently random imposition of variations which may have little connection to the local character—e.g. where different roofing materials are used on a terrace of houses. The concept of the street segment was borrowed from Space Syntax where it is defined as the section of axial line or street or path lying between two intersections (Space Syntax Online Training Platform 2020). The segments approximate to the experience of the user and the

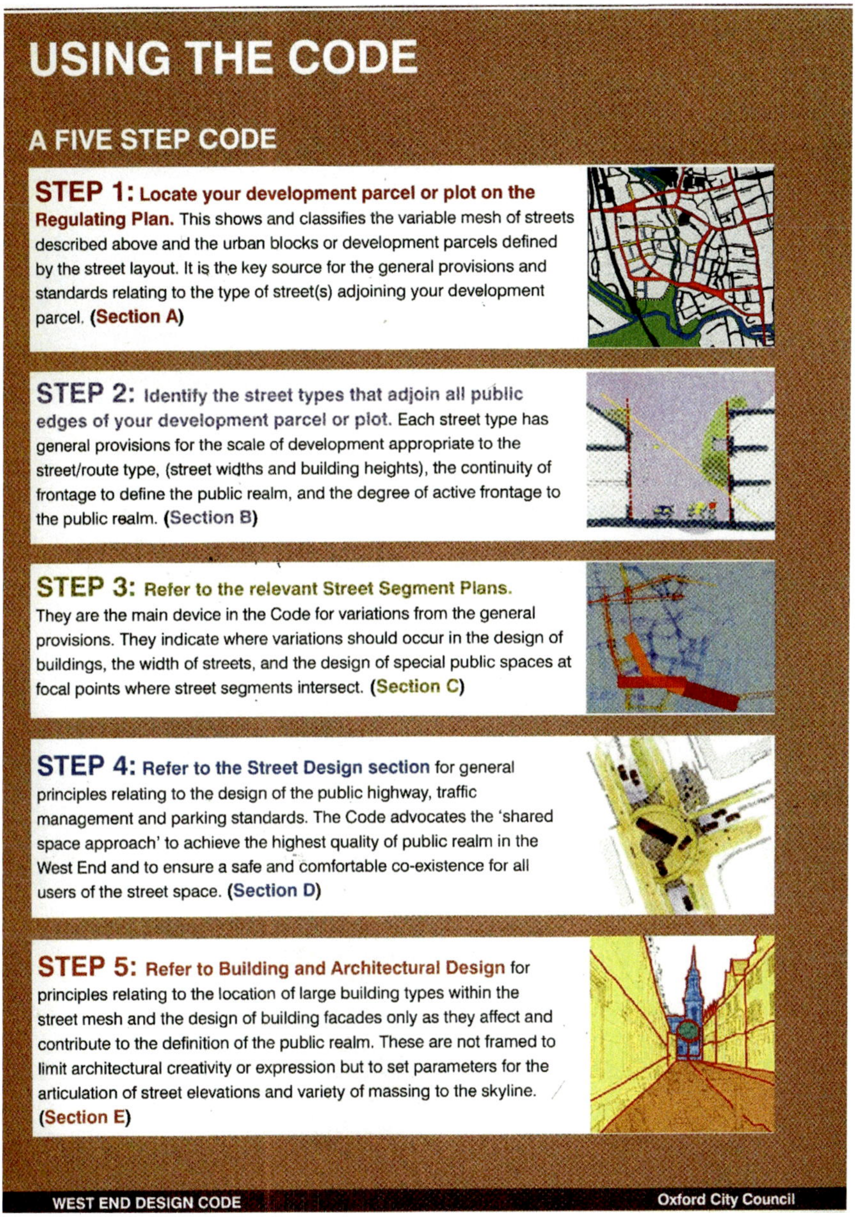

USING THE CODE

A FIVE STEP CODE

STEP 1: Locate your development parcel or plot on the Regulating Plan. This shows and classifies the variable mesh of streets described above and the urban blocks or development parcels defined by the street layout. It is the key source for the general provisions and standards relating to the type of street(s) adjoining your development parcel. **(Section A)**

STEP 2: Identify the street types that adjoin all public edges of your development parcel or plot. Each street type has general provisions for the scale of development appropriate to the street/route type, (street widths and building heights), the continuity of frontage to define the public realm, and the degree of active frontage to the public realm. **(Section B)**

STEP 3: Refer to the relevant Street Segment Plans. They are the main device in the Code for variations from the general provisions. They indicate where variations should occur in the design of buildings, the width of streets, and the design of special public spaces at focal points where street segments intersect. **(Section C)**

STEP 4: Refer to the Street Design section for general principles relating to the design of the public highway, traffic management and parking standards. The Code advocates the 'shared space approach' to achieve the highest quality of public realm in the West End and to ensure a safe and comfortable co-existence for all users of the street space. **(Section D)**

STEP 5: Refer to Building and Architectural Design for principles relating to the location of large building types within the street mesh and the design of building facades only as they affect and contribute to the definition of the public realm. These are not framed to limit architectural creativity or expression but to set parameters for the articulation of street elevations and variety of massing to the skyline. **(Section E)**

WEST END DESIGN CODE Oxford City Council

Fig. 4.7 OWEDC, a guide to the use of the Design Code (*Source* Oxford City Council 2008)

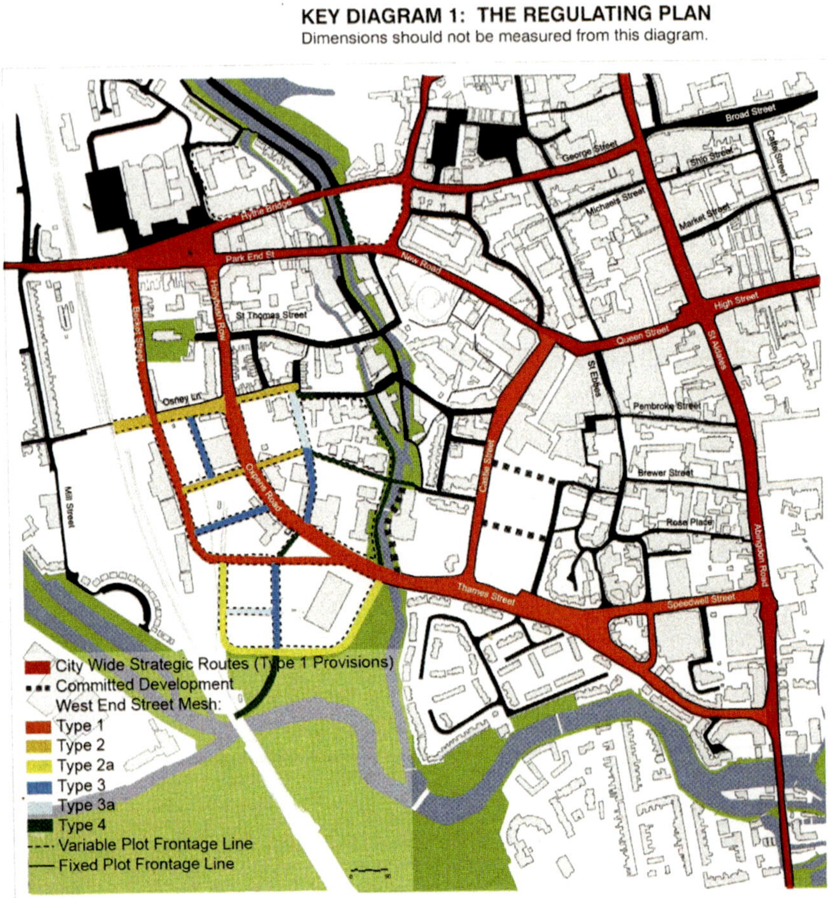

Fig. 4.8 OWEDC, the Regulating Plan (*Source* Oxford City Council 2008)

intention in this project was to use them as a way of indicating where changes in building form or horizontal surface treatment should occur (Figs. 4.11 and 4.12).

Section D (Step 4). Advised by Ben Hamilton-Baillie Associates, the redevelopment of the West End was regarded as an opportunity to introduce an innovative approach to the design and management of the street system. This moved away from standards dominated by a concern to maximise efficiency in traffic movement to proposals which, based on the street segment plan, would reinforce the visual distinction between different parts of the network. A limit to carriageway widths was introduced and priority was given to pedestrians, cyclists and public transport users. These proposals supported the recently introduced Manual for Streets (Department of Transport 2007) which was mainly directed towards residential areas. They also derived from the experience of the Dutch *woonerf* and, as an exception to the practice

SECTION B: TYPE ONE 27

Type One: Mixed Use Main Street

Scale		Continuity Building Frontage		Active Frontage
Range of Widths for Public Highway in metres	Range of Building Heights measured in storeys	Building Line (BL)	Boundary Treatment (BT)	Grade Frontage (doors/100m) Transparency facade (%)
Minimum 15m Maximum 27m Measured between Plot Frontage Lines	3, 3.5, 4, 4.5 and 5 Storeys Ground floor to floor minimum 3.5m	BL/PFL coincide (0-1m) for 95-100% of Plot Width Remaining 5% may be setback or break in building frontage Breaks/gaps max. 5m	Where break in BF - railings, walls, planting mark PFL	A, B, C Grade Frontage - 6-20 doors/100m Transparency ≥ 30%

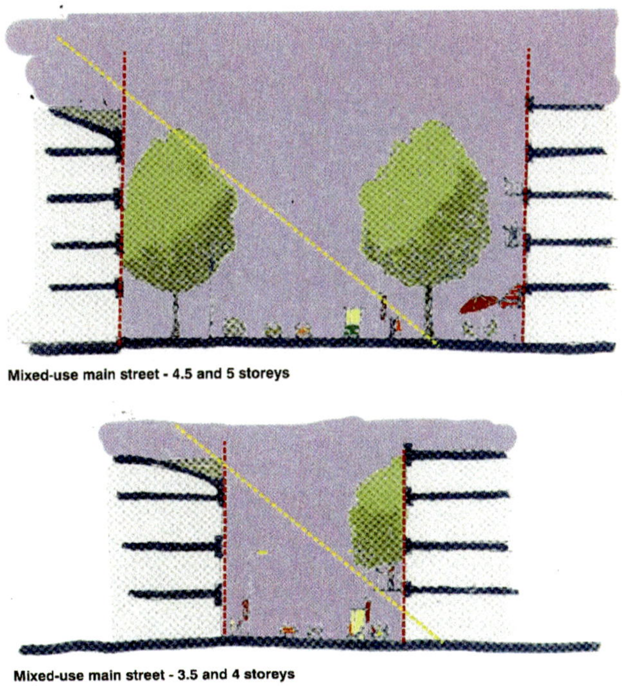

Fig. 4.9 OWEDC, alternative sections, scale and frontage instructions for Type One Street (*Source* Oxford City Council 2008)

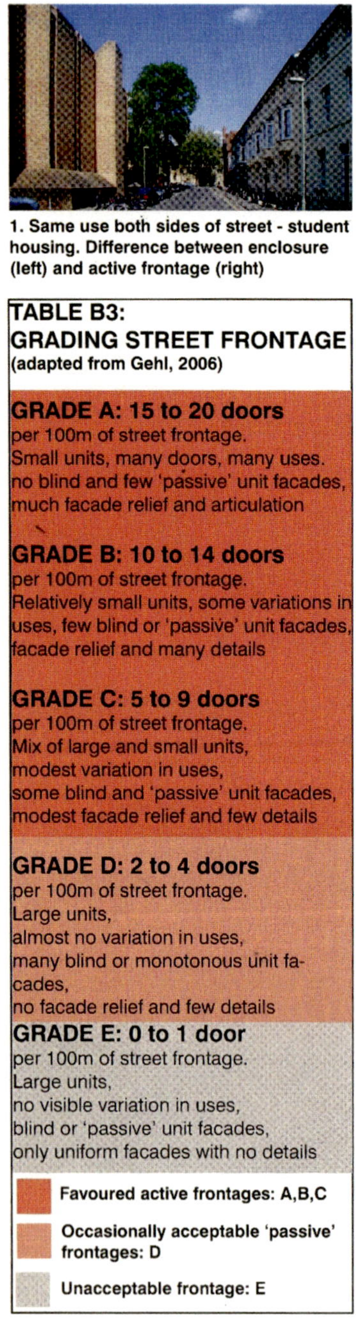

Fig. 4.10 OWEDC, grading of street frontage activity (*Source* Oxford City Council 2008)

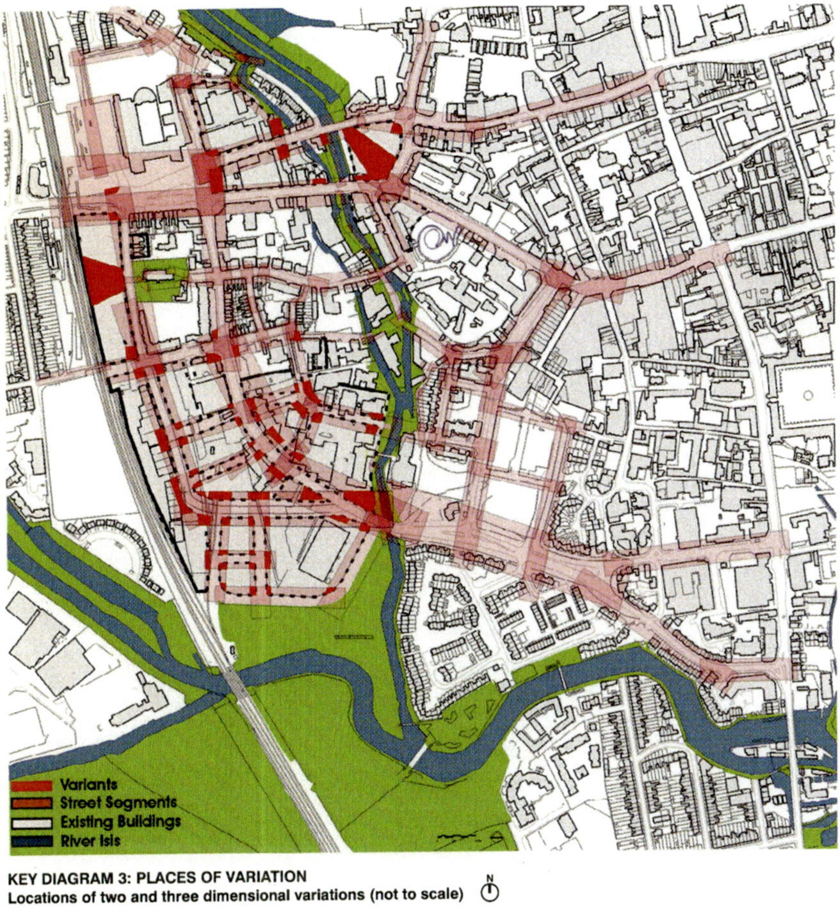

Fig. 4.11 OWEDC, street segment plan (*Source* Oxford City Council 2008)

in the rest of the code, the examples shown in this section were drawn from Dutch and German cases because of the absence of local examples.

Section E (Step 5). This was deliberately the shortest section of the code because this was the aspect of the Asnières POS which was most heavily criticised and attempts to influence building design through comprehensive and detailed prescriptions are often regarded as inhibiting design creativity. Therefore, the architectural idiom was not restricted but some principles regarding their impact on the public realm were introduced. These included linking the degree of richness and articulation of facades according to viewing distances—i.e. the greatest variation should occur at ground level from where the building is most viewed (Gehl et al. 2006). There was also a requirement to articulate building silhouettes when seen both from

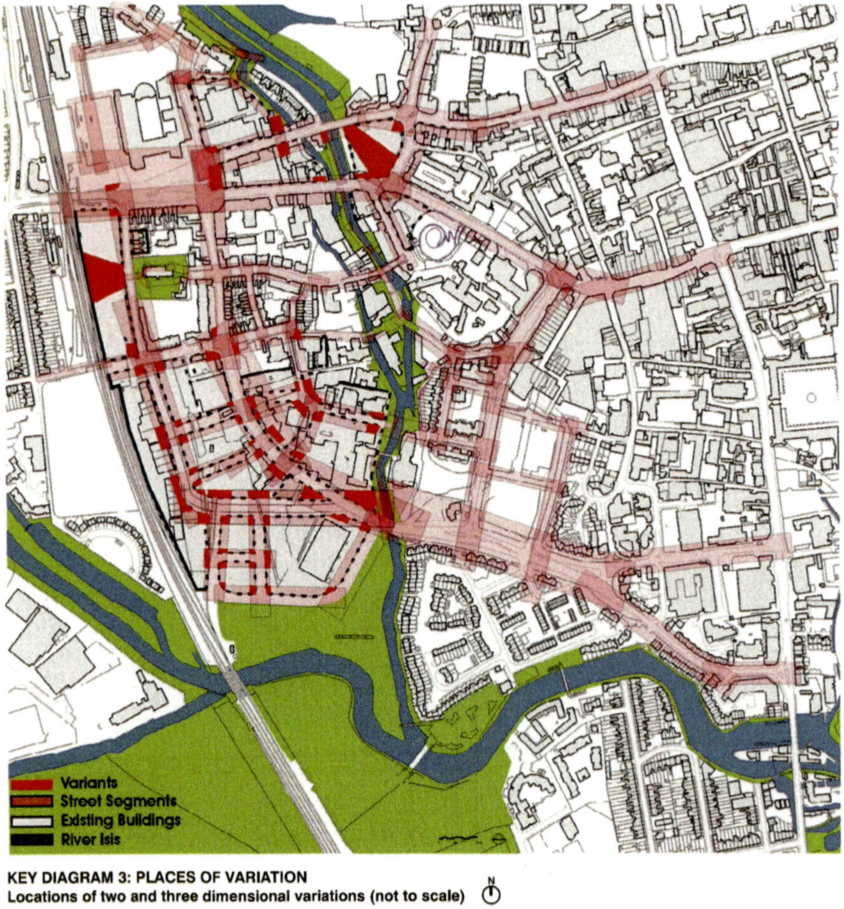

Fig. 4.12 OWEDC, location of variations to building form and/or horizontal surfaces (*Source* Oxford City Council 2008)

the street and from a longer distance. These points were illustrated by recent buildings in Oxford in order to emphasise that it was not necessary to return to past styles to achieve these qualities.

The design code was tested at different stages of the plan preparation and through a design exercise with development control officers, local architects and a representative of the Commission for Architecture and the Built Environment (CABE). The Planning Inspector who reviewed all the parts of the West End Action Plan, very succinctly summarised the intention of the authors of the Design Code as being 'a comprehensive document based on a combination of general urban design principles and the place specific qualities of Oxford City Centre and the West End. Its priority is to set the relationships between building facades and the public realm, the form and mass of buildings in relation to one another, and the scale and types of streets and urban blocks. It has sought to identify the least number of most significant and long – lasting elements of the public realm of the West End in order to provide a flexible framework for the generation of a new, successful and highly locally distinctive public realm (…) I consider that this innovative design code (…) will ensure that local features that make the area distinctive are considered and built upon in its renaissance' (Bussey 2008, para. 5.3.3).

4.5 The Application of the West Oxford Design Code

Unfortunately, the code was adopted the same year that the financial crash of 2008 halted development, especially in the commercial field. The priority of the City Council was to achieve the completion of a project to renew and extend the Westgate shopping centre built in the 1970s and sited within the area of the Action Plan. Apart from the incorporation of two open pedestrian routes crossing the new centre which had been included in the Regulating Plan, any reference to the design of the new centre was deliberately excluded from the provisions of the code. This was in order to facilitate its completion because it had been subject to a long planning process of application, rejection, appeal and resubmission which started in 1998. It was finally approved in detail by the Council in 2014 and completed in 2017. It is particularly notorious for the way the code provisions for active frontages on public streets and sleeving large buildings with smaller plots were not considered in the design as executed.

The OWEDC was intended to remain in force from 2008 until 2016. In fact, it was only in 2020 that the City Council announced its intention to prepare a new West End plan. Although no systematic study of its implementation has been made, anecdotal evidence from discussions with City Council planning officers indicates that it continues to play a role in directing and justifying their discussion on proposals for the area even though few developments have taken took place.

In 2013 a Master Plan for the largest potential development site in the West End was adopted (Oxford City Council 2013), however, to date no development according to this plan has yet taken place. This Plan shows the optional road mesh of the Code adopted in its essentials because of the extent of the residential development which was proposed. But the attempt to ensure a successful relationship between public and private space, without specifying architectural detail, had been replaced by such anodyne statements as the buildings 'should be of high architectural design quality' (para. 4.4) or 'proposals should seek to enhance the quality of the public realm for all' (para. 4.6) or 'high quality architecture will be expected in all new development' (para. 5.6).

4.6 Conclusions

Although these plans were prepared decades ago and planning legislation in both countries has changed as have the instruments available to control, protect and enhance the built environment, there may still be lessons to be learnt from these very diverse cases. The first is the need to revisit plans, guides and codes to assess the way they have been used and aspects of their preparation which were successful and could be repeated elsewhere and on the contrary those elements which were problematic and should be used with care if at all. To paraphrase Stewart Brand (1994) a planning instrument's life begins after its authors have delivered it to the implementing authority. This is clearly most relevant where a code is prepared by external consultants, as in the above two cases, but even if a code is prepared internally by a public agency, staff changes can result in a different team being responsible for its implementation. This might seem obvious, but this type of investigation is rarely undertaken and there seems to be a parallel with architects anxious to celebrate their buildings in publications before they are occupied.

A second lesson of these two cases is the advantage of integrating student academic projects with real-life consultancy work undertaking plan preparation where this is labour-intensive as in the Asnières case, and retrospective evaluations. In the academic tradition of the architecture studio, the project is an instrument for exploring problems and indicating, if not providing, a range of solutions. There is no single output as might be expected in a scientific project and this flexibility replicates the real-life situation of consultancy. It would not have been possible to execute the Asnières POS without the unpaid input by postgraduate students. The experience of Qualipos demonstrated how difficult it was to make this type of project financially viable. Although the OWEDC was funded by the City Council, the work was subsidised to the extent that there were no overheads such as office accommodation costs to be met by the ad hoc team assembled for the project.

Another important lesson from Asnières that was carried forward to the OWEDC was that in preparing the instrument we not only had to consider that those responsible for its implementation will not be the authors, but even if, initially, they are responsible, staff change whether for administrative or financial reasons. At Asnières this change rendered the POS inoperable. The OWEDC was therefore organised so that the process of understanding its main arguments was very straightforward to all those who might be engaged in its future implementation.

Both the POS and the OWEDC received plaudits from outside professionals, the former from the notaries and the latter from an independent government planning inspector. However, this did not ensure their ability to influence future building form. In the latter case the wider economic circumstances, landownership questions and the various political interests in the context of a complex city, hindered the process of development so that eventually, it was replaced by other plans. In both cases attempts to influence the detailed form of buildings proved unsuccessful. In the first case the proposals were too detailed and too expensive to implement. In the latter case, even though the broader proposals were carried forward into a successor plan, the regulations concerning the relationship of new buildings with the public realm were ignored. However, the fact that some aspects of each instrument were enduring suggests that the way the argument for the proposals had been developed, based on systematic investigations of urban form, had a useful didactic function which outlasted the POS and the OWEDC.

Anne Moudon, the first President of ISUF, wrote in 1994 that we may envy the thoroughness of the studies of some of our colleagues, but we often do not have the possibility either from a legislative or resource viewpoint to undertake deep morphological investigations in practice. The financial deficits experienced by the practice set up after the Asnières POS and the limitations of attempts to impose rigorous rules on building form in both the cases described in this chapter, clearly confirm her observation. If we are to penetrate the silo of practice then the challenge we have is to adopt morphological practices that incorporate both analytical and prescriptive elements which are easy to implement, which are not prohibitively resource-intensive and which respond to the legislative context.

References

Barke M (2015) Further thoughts on research and practice in urban morphology: a British perspective. Urban Morphol 19:96–99

Bentley I, Alcock A, Murrain P, McGlynn S, Smith G (1985) Responsive environments, a manual for designers. Architectural Press, Oxford

Brand S (1994) How buildings learn, what happens after they are built. Viking, New York

Bussey S (2008) Inspector's report on West End Area Action Plan. Oxford City Council. https://www.oxford.gov.uk/downloads/download/422/west_end_area_action_plan. Accessed 14 Apr 2020

Cullen G (1961) Townscape. Architectural Press, London

Department of the Environment Transport and the Regions (DETR) (2000) By Design. Her Majesty's Stationary Office, London

Department of Transport (2007) Manual for streets. Her Majesty's Stationery Office, London http://www.dft.gov.uk/pgr/sustainable/manforstreets/. Accessed 14 Apr 2020

Essex County Council (1973) Essex Design Guide. www.essexdesignguide.co.uk/about/history-of-the-essex-design-guide/Accessed 2 Apr 2020

Form-Based Codes Institute (2020) Five main elements of form-based codes https://formbasedcodes.org/definition. Accessed 28 Apr 2020

Gautier P, Gilliland J (2006) Mapping urban morphology: a classification scheme for interpreting contributions to the study of urban form. Urban Morphol 10:41–50

Gehl J, Johansen Kaefer L, Reigstad S (2006) Close encounters with buildings. Urban Des Int 11:29–47

HRH The Prince of Wales (1989) A vision of Britain; a personal view of architecture. Guild Publishing, London

Joint Centre for Urban Design Oxford Polytechnic (1980) A Report on the Ville Basse of the City of Provins, Unpubished Report for the Patrimoine Historique et Artistique de la France rot

Kropf K (1993) An enquiry into the definition of built form in urban morphology. Unpublished PhD thesis University of Birmingham, UK

Larkham PJ (2006) The study of urban form in Great Britain. Urban Morphol 10:114–171

Larice M, MacDonald E (2007) The urban design reader. Routledge, Abingdon

Lynch K (1960) The image of the city. MIT Press, Cambridge

Mairie d'Asnieres sur Oise, Samuels I, Kropf K, leurs Collaborateurs (1992) Plan d' Occupation des Sols. Commune d' Asnieres sur Oise, Asnieres sur Oise
Mairie d'Asnieres sur Oise (1992) Lettre d' information de la Commune d'Asnieres -sur-Oise. Commune d'Asnieres-sur-Oise, Asnieres sur Oise
McGlynn S (1993) Reviewing the rhetoric. In: McGlynn S, Hayward R (eds) Making better places. Butterworth-Heinemann, Oxford
McGlynn S, Samuels I (2000) The funnel, the sieve and the template: towards an operational urban morphology. Urban Morphol 4:78–89
Merlin P, Choay F (1988) Dictionaire de l'urbanisme et de l'amenagement. Presses Universitaires de France, Paris
Moudon AV (1994) Getting to know the built landscape: typomorphology. In: Frank KA, Schneekloth L (eds) Ordering space types in urban design. Van Nostrand Reinhold, New York
Nairn I (1956) Outrage. Architectural Press London (reprint of Architectural Review June 1955)
Oxford City Council (2008) West End Area Action Plan. Oxford City Council, Oxford https://www.oxford.gov.uk/downloads/download/422/west_end_area_action_plan. Accessed 14 Apr 2020
Oxford City Council (2013) Oxpens, Oxford West End Masterplan Supplementary Planning Document, Oxford City Council, Oxford https://www.oxford.gov.uk/downloads/download/153/oxpens_masterplan_spd. Accessed 14 Apr 2020
Pattacini L (1996) Better places to live in; learning from the French experience. Unpublished MA thesis, Oxford Polytechnic, UK
Placemaking Associates, McGlynn S, Samuels I, Stanley M, LeRoux M, Matumoto A (2007) West End Area Action Plan, Appendix 3, West End Design Code. Oxford City Council, Oxford. https://www.oxford.gov.uk/downloads/file/1434/west_end_aap_volume_2_-_appendices. Accessed 20 March 2020
Rapoport A (1977) Human aspects of urban form; towards a man-environment approach to urban form and design. Pergamon Press, Oxford
Reps J (2016) Bastides, medieval new towns of southwestern France. bastides.library/cornell.edu/about. Accessed 20 March 2020
Samuels I (1993) The Plan d' Occupation des Sols for Asnières sur Oise: a morphological design guide. In: McGlynn S, Hayward R (eds) Making better places. Butterworth-Heinemann, Oxford
Samuels I (1999) A typomorphological approach to design: the plan for St Gervais. Urban Des Int 4:129–141
Samuels I (2020) Poundbury revisited. U+D Urban Form Des 13:62–65
Samuels I, Oliver P (1987) Donzenac: recomendatins for its future. Unpublished report for the Parimoine Historique et Atistique de la France
Samuels I, Pattacini L (1997) From description to prescription: reflections on the use of a morphological approach in design guidance. Urban Des Int 2:81–91
Scheer B (2008) Urban morphology and urban design. Urban Morphol 12:140–142
Space Syntax Online Training Platform (2020) Segment http://otp.spacesyntax.net/?s=street+segment+&post_type=term. Accessed 28 Apr 2020
Worskett R (1969) The character of towns, an approach to conservation. Architectural Press, London

Chapter 5
Is There a Normative Science of the Built Environment?

Karl Kropf

Abstract There has been a long-running debate within urban morphology around the 'description-prescription' problem. The central question is whether we can derive prescriptions for new development based on descriptions of existing and historic development. The debate is sharpened when we seek to make the descriptions provided by urban morphology more objective and scientific with the expectation that an objective, scientific description should not, in principle, be normative. This chapter continues the debate by taking up the idea of normative science as introduced by CS Peirce and extended by JJ Liszka. In brief, Peirce's notion focuses on the relationship between human purposes and the performance of our constructions in seeking to achieve those purposes. In exploring how the idea of normative science might help build the bridge between urban morphology and planning and urban design practice, the chapter points to the importance of teaching and asks, do we necessarily operate within the realm of ethics?

Keywords Description · Prescription · Normative science · Ethics · Purpose · Performance

> The classical pragmatists, Peirce, James, Dewey and Mead, all held that value and normativity permeate all of experience. In the philosophy of science, what this point of view implied is that normative judgments are essential to the practice of science itself. (Hilary Putnam 2002, p. 30)

5.1 Introduction

The notions of value and normativity are central to the topic of this volume in a number of ways. The brief to the authors, as should be clear, was to state a morphological view on the processes of city building and to provide an example of how that view is expressed in design. To give a view and illustrate its expression as a design is to make

K. Kropf (✉)
Oxford Brookes University, Oxford, UK
e-mail: kkropf@brookes.ac.uk

V. Oliveira (ed.), *Morphological Research in Planning, Urban Design and Architecture*, The Urban Book Series, https://doi.org/10.1007/978-3-030-66460-2_5

a statement of values. Design is a fundamentally normative practice. We put forward what we think ought to happen based on a set of implicit or explicit values both about the design itself and the way that it was generated. What the quote from Hilary Putnam above suggests, however, is that values and norms are just as relevant to another topic of this book: that, 'urban morphology aims at scientifically describing the physical form of cities [and] explaining how different agents and processes shape this form over time…'. Putnam's point, and that of the authors he cites, is not that the permeation of science with values and normativity therefore renders it hopelessly subjective but that we need to be clear that science is built on a set of shared values. We can then be clear when and where we are exercising our values and, ideally, openly express them.

To that end and to provide an initial point of reference, it is worth providing a kind of position statement for this chapter in relation to the overarching theme of the book. In particular, it relates to the questions of whether urban morphology is a scientific, 'objective' endeavour and whether it can, at the same time, be a basis for design principles.

Urban morphology should not be 'for' or 'against' any particular period, principle or style of architecture or urbanism—even if some important contributors to the field are or have been motivated by the perceived failures of twentieth-century Modernist architecture and urbanism. If urban morphology is to be scientific and objective, which I believe it should, it cannot take sides. It has to remain objective and describe and explain all forms created and the processes that create them—including the 'ruptures' that are common to all natural history. Once we have done that, then we can make judgements, not as urban morphologists per se, but as individual architects and urbanists working in a particular time and context. This situation is, in my view, one of the central reasons that the 'bridge between morphological knowledge and action on the city has never been made robust'. If urban morphology 'as a discipline' advocates for a particular principle or style 'and' claims objectivity, there is a deep-seated, internal contradiction. We cannot have it both ways. This is not to say we cannot use objective, scientific methods to determine how particular forms and configurations perform with regard to particular social and professional aims and preferences but we can only do that 'objectively' by being open and clear about the particular social aims and preferences—which may or may not be shared across a population. Indeed, one of the criticisms of Modernism was that its principles were claimed to be Universal.

5.2 The Description-Prescription Problem

The general background to these issues is, of course, that urban morphology has been developed, particularly in Italy and France, by architects and urbanists to inform their designs in response to the perceived crisis of twentieth-century Modernism. With progressive attempts to apply urban morphology to planning and urban design in a more systematic way (see, for example, Samuels and Pattacini 1997; Kropf 2001; Oliveira et al. 2014) a debate began to emerge around the relationship between description and prescription. Can design prescriptions that generate 'good form' be derived from systematic or scientific descriptions and if so, on what basis? A further question might be whether such principles could be universal.

The common objection to claims that design principles, universal or not, can be scientifically derived goes something like this.

There is nothing in the existing state of cities that determines how they either will or should be built in the future. As much as we might study cities, they cannot 'tell us' what to do next. We always have to make a choice. This is the description-prescription problem, which is a version of the is-ought problem in philosophy as identified by David Hume (1964), sometimes referred to as the naturalistic fallacy.

Even if we consider cities as a 'quasi-natural' phenomenon, there is no logical certainty in moving from observations to predictions due to the problem of induction as again identified by Hume (1964). As formalised by C.S. Peirce, the move from observation to prediction is necessarily made by 'abduction' or hypothesis but remains uncertain. As further developed by Popper, any hypothesis can only be refuted and not 'proven' by subsequent induction.

And such a shift to the 'natural' perspective, even quasi-natural, can contribute to the toughness of the description-prescription chestnut. The fact that cities and the societies that build and live in them have been the subject of various fields within the social sciences for many years does not resolve the description-prescription problem. Nor does the emerging notion of a 'science of cities' (Batty 2013; Mehaffy 2014a, b; Wilson 2012). However much the development, growth and transformation of cities can be seen and modelled as complex adaptive systems and a problem of organised complexity, cities remain, inescapably, human constructions based on human choices. For that reason, cities sit firmly in the realm of the is-ought problem and so the realm of ethics. But in stating that point so bluntly, objections immediately begin to arise. So, perhaps the toughness of the chestnut is a function of the categories rather than the phenomenon itself. We seem to be unduly bound by thinking of urban morphology as either a purely 'objective' discipline or a matter of preference, design and ethics.

This issue is at heart of the point made by Putnam when citing the classical pragmatists. The quote is from a critique by Putnam of the fact/value dichotomy that puts facts on one side and values on the other and never the twain shall meet. Putnam argues that the strict dichotomy is a false one, but as he goes on to state, the acknowledgement that facts are permeated by values does not result in a distinction less soup of subjectivity. Rather, it clarifies that there are distinct sets of values. There are, on the one hand, 'epistemic' values and principles that define what we consider to be good descriptions of the world and, on the other hand, there are ethical and aesthetic values. The planning and design of cities fall within the ethical realm on the basis that planning and design are predicated on what is 'good' from a human perspective and involves human action that in some degree requires justification within a group.

That is to say, we use different sets of values for deciding: (i) what is a good, objective description of urban form, and (ii) what are good forms to build and use in particular circumstances for particular purposes. There are different criteria we apply in determining what is an accurate description of urban form and which forms might work best for a given human activity which begs the question as to whether there might be a 'scientific' basis for deciding which forms to select for a given project. As it happens, one of the pragmatists cited by Putnam, C.S. Peirce, put forward the concept of Normative Science.

> Normative science considers the phenomenon only so far as it can be controlled, compares purpose with performance, and ascertains the general principles of the relation between them. (Peirce 1983)

5.3 The Idea of a Normative Science of Ethics

As a basic notion, the idea of Normative Science would seem to offer a way over the apparent impasse of the description-prescription problem. The first step is to openly acknowledge the normative basis of the endeavour, the second is to focus on purpose, performance and the relation between them. The obvious question is then how such a normative science would work in more detail and in particular a normative science of the built environment. Peirce himself did not fully elaborate the idea but a further exploration of the implications of a normative science of ethics has been set out by the philosopher James Jakób Liszka (2014), also drawing on the work of John Dewey.

A key starting point in Liszka's account is the Peircean conception of purposive, goal-directed behaviour. 'If normative claims are about what ought to be done, then for Peirce, the word ought has no meaning except relatively to an end. That ought to be done which is conducive to a certain end. Given an end, one can measure whether actions have either achieved it or not' (2014, p. 463).

Liszka then concludes that 'the basic elements of purposive behaviour are the end desired, the means believed to attain that end, and the actions taken in that regard' (2014, p. 463).

Going further, Liszka points out that there are two types of 'normative claim': (i) claims about which is the best means to achieve a particular end, and (ii) claims about which of the different ends is best to achieve?

The need for the distinction arises because there are many ends that might be pursued, some of which may overlap or conflict.

> An ideal of conduct or an end worthy of pursuit is one that is consistent with other ends or ideals considered worthy of pursuit Negative approach to good results from the reduction of faults, limitations, and problems with current practices… The pursuit of any end is always within a network or system of ends which are also being pursued. (2014, p. 473)

A further consequence arises with the acknowledgement of a plurality of potentially conflicting ends, in particular when the aim is to achieve a 'reasonable' overall position. Seeking such a position forces one to generalise the ends to a point where the 'ultimate good' (*summum bonum*) cannot be fully articulated in advance. Rather, the result is to shift from a 'positive' approach to a 'negative' one in which, 'there is no predetermined ultimate end, but good results from the reduction of faults, limitations, and problems with current practices' (Liszka 2014, p. 474).

As Liszka points out, Peirce conceived of this as a process of 'evolution'. Characterising the point further, he quotes Beverley Kent, '[Peirce] refused to grant that the aesthetic ideal must be a static result. By admitting process, Peirce was no longer limited to a self–satisfied ideal. He could adopt an end that would always anticipate an improvement in its results (154). The goal in this case is not to perfect an ideal, but to make things better' (Liszka 2014, p. 475).

In adopting a negative approach of 'making things better' the focus shifts to a process of identifying problems and generating hypotheses about how best to solve them. In making that shift the process also becomes one that satisfies the criteria of scientific enquiry.

> Making things better, intuitively speaking, involves making things that are problematic less so. That involves solutions to those problems. Proposed solutions to problems are, in fact, abductions – in Peirce's language – hypotheses about changes in practices that will result in ameliorating the problem. Just as in scientific practice, problems and anomalies with existing hypotheses become the impetus for their resolution by new hypotheses that are then inductively tested, so proposed solutions to existing practices become practical hypotheses that could, in principle, be subject to testing. (Liszka 2014, p. 475)

More particularly, citing Dewey, Liszka adds that 'In the context of means-end conduct, there are two types of inquiry to consider, inquiry into the lacks, conflicts, and problems of the existing situation, and second, inquiry into whether the means hypothesized will resolve those problems and satisfy the lacks. Using Peirce's language, abduction is the reasoning involved in the first sort of inquiry, while deduction and induction, the reasoning involved in the second sort. Thus, in principle, normative issues are subject to the same sort of reasoning as scientific ones' (Liszka 2014, p. 476).

Going back to Peirce's generalised characterisation of normative science as involving purposes, performance and the relation between the two, it is possible to paraphrase and elaborate the above quote. The purpose, in Peirce's terms, is to resolve or ameliorate the conflict or problem and the performance is the extent to which a means hypothesised to resolve the problems is satisfactory. The relationship between the two is a two-step process.

The first step involves: the 'irritation' of a problem or problems and the identification of a purpose; a survey and assessment of the issues involved in the problem to refine the purpose; a search for possible solutions; the selection and putting forward of a solution as a means to resolve the problem (abduction or hypothesis); mentally (or by models and simulations) working through the potential implications of the hypothesised solution and making adjustments as necessary (deduction); comparing the proposed solution against established measures and/or other successful examples to check for anticipated performance and potential issues and making further adjustments as necessary (induction). The deductive and inductive steps are checks on the performance of the proposed solution in principle.

The second step involves: selecting the final 'design' for the solution (hypothesis/abduction); putting the solution into action (e.g. constructing a building); working through any issues arising at the outset such as unresolved details, implementing corrections as necessary (deduction); putting the solution into full operation; assessing its performance on the basis of ongoing experience in comparison with the initial aims and objectives (induction).

Finally, Liszka raises a point about the operation of normative claims or rules within the context of practical reasoning. In simple terms, practical reasoning involves the desire or intention to achieve some end, the belief that undertaking a particular course of action will achieve the end and the decision to undertake the action. What Liszka points out is that the 'schema' of practical reasoning provides a basis for both explanation and justification for actions.

To paraphrase (2014, p. 469), practical reasoning provides an explanation for action by laying out the motivation of an action, which is the belief that the action will achieve the aim. Whether it is a true belief or not does not alter the explanation but it does affect its justification. Following a mistaken belief still explains an action even if it does not justify it. Once a given action has been established to lead successfully to a particular end, it can then be used as a justification for what 'ought' to be done if that is the desired end. It can become a normative 'rule' (though always subject to further meta-rule that success may be contingent on a wide range of variables that may not always be in place or change over time).

In summary, the key elements of a framework for normative science are that: (i) it is conceived as an ongoing, evolutionary process; (ii) the process is prompted or invoked in response to a perceived 'problem'; (iii) there are three main elements that come into play in response to a problem: the desired end or purpose (a resolution of the problem), the means believed or claimed to achieve the end and the action; (iv) there are two steps in the process of making a normative claim to serve a purpose and checking its performance, both of which involve abduction, deduction and induction; (v) there are (at least) two types of normative claim: those about the best means to achieve a given purpose and those about the best purposes to achieve; and (vi), for a given purpose, the justification for a normative claim or 'rule' is the evidence that the means is likely to achieve the purpose.

5.4 Normative Science and the Typological Process

In getting to this point, and as a kind of provocation, it was suggested that because cities remain inescapably the result of human choices, any normative science of cities must sit firmly in the realm of ethics. In many ways this should not be controversial on the simple basis that decisions about where and how to build cities can have profound qualitative effects on many peoples' lives. It should therefore be equally uncontroversial that a normative science of the built environment might be based on a framework for a normative science of ethics.

Another reason for adopting such a framework as a working hypothesis is the striking correspondence between the process summarised above and the typological process as put forward within urban morphology (Caniggia and Maffei 2001; Kropf 2017; Oliveira 2016; Petruccioli 2007).

The most obvious and important points of correspondence are as follows.

Both are conceived as evolutionary processes. The prompt for invoking the process in both cases is an 'irritation' or disturbance, perceived as 'problems' in the Framework for a Normative Science of Ethics (FNSE) and a 'crisis' in the typological process.

Both involve three main elements. In the case of the FNSE these are: (i) the desired end or purpose, (ii) the means believed to achieve the end and iii. the action. For the typological process the elements are (i) a need/desire and intention to build, (ii) the idea of the proposed building (the a priori type) believed to satisfy the intention to build and (iii) the act of constructing and using the building (Fig. 5.1a).

Both involve a two-step process in making a normative claim and checking its performance. In the FNSE, there is: (a) an initial consideration of the options and mental testing, and (b) putting the selected solution into action and reflecting on its performance against the anticipated result. In the typological process there is: (a) the design stage, in which options are considered and tested 'on paper', and (b) constructing the design, putting it to active use and assessing the result against the anticipated outcome and/or some criteria for success based on the initial problem. Both steps in both cases involve abduction/hypothesis, deduction and induction.

For both the typological process and FNSE, there are two types of normative claim or hypothesis. For the FNSE, there is: (i) the claim about the best means to achieve a particular end, and (ii) the claim about the best ends to pursue. In the case of the typological process, the corresponding types of claim are: (i) the choice of the best design or type to solve a given design problem, and (ii) the choice of which design problems are the most important to address (Fig. 5.1b). The difference between these two types in both cases also corresponds to a distinction in more general models of evolutionary processes between i. feedback or somatic change, and ii. calibration or genetic change. The distinction is explored in more detail in Kropf (2014), but a key point relevant to the idea of a framework for a normative science of ethics is that, in general, feedback operates at the level of individual agents while calibration operates at the level of the population. An example that bears on both ethics and the built environment is the human response to climate change. The 'irritation' induced by evidence of climate change is prompting a crisis that is forcing a debate within the wider population about the most important aims to pursue in the design and construction of the built environment. The result is a progressive 'recalibration' of norms in terms of the range of professionally and socially supported goals. With that recalibration and the identification, for example, of carbon reduction as a socially sanctioned goal, the second type of normative claim comes into play on individual projects. The question becomes, which designs are the best at achieving the goal of carbon reduction? The purposive norms are then developed through a process of feedback by testing the performance of hypothesised designs.

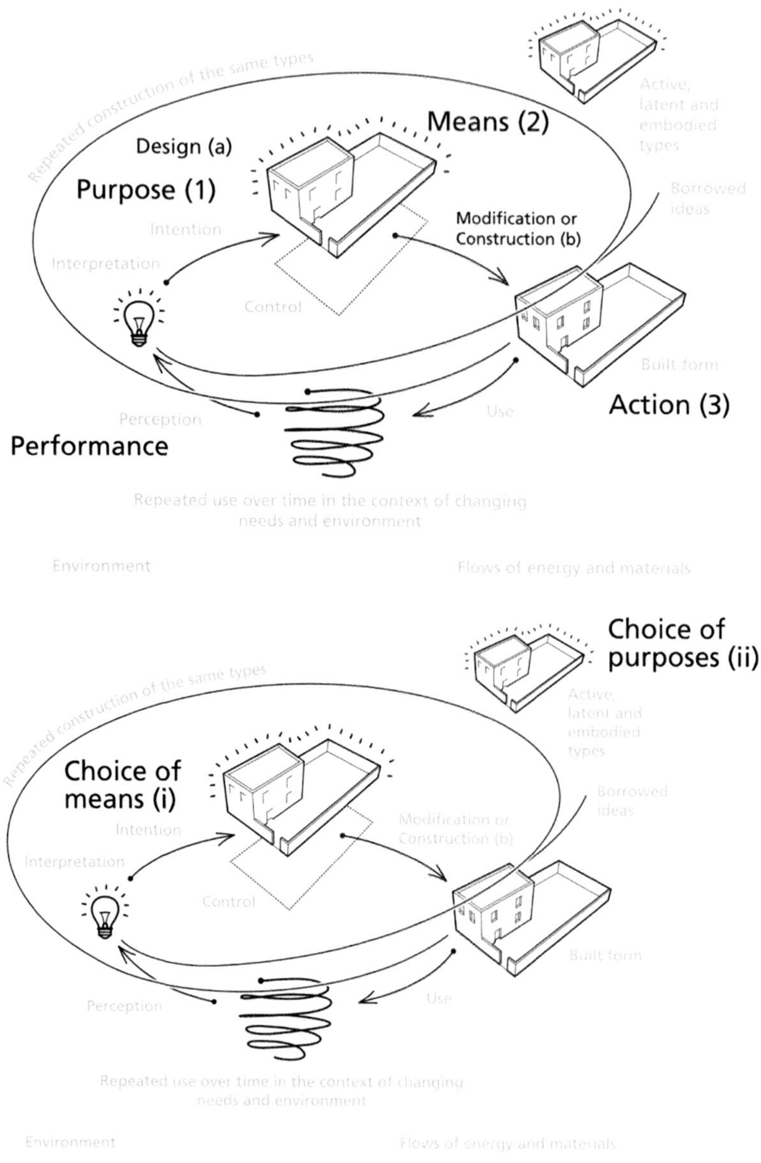

Fig. 5.1 Diagram of the typological process, highlighting the common elements with Liszka's Framework for a Normative Science of Ethics. **1** Purpose, **2** Means, **3** Action; **a** Design/mental testing, **b** Modification/construction (top). **i** Choice of means/feedback, **ii** Choice of purposes/calibration (bottom)

5.5 The Potential Role of Urban Morphology in a Normative Science of the Built Environment

If the idea of a normative science of the built environment has some coherence and potential benefit as an initial hypothesis, the next question is, what is the role of urban morphology within that framework? Given the limited scope of this chapter, the following presents only a summary exploration of the possibilities framed in very general terms.

At the most general level, the question takes us back to the distinction discussed at the outset between the different sets of values we use for deciding i. what is a good, objective description of urban form, and ii. what are good forms to build and use in particular circumstances for particular purposes. One purpose is to provide an accurate description, the other is to provide a built form to accommodate a particular activity or, at its most general, 'good urban form'. As set out in the 'position statement' at the beginning of the chapter, in my view, the primary aim of urban morphology is to provide a comprehensive and rigorous description of the built environment. It follows, in my view, that whatever the specific means adopted, the best result is most likely to come from limiting the goal to description and not including the identification of 'good urban form'. Making judgements about whether a form is good or bad requires a different and more specific purpose in order to establish the performance criteria for success. It would also seem that having an accurate, objective description would provide the most secure basis for making judgements about the performance of forms against criteria for a given purpose and to understand how the form achieves the aim. That is, accurate descriptions, free of judgements, would seem to be a precursor to understanding how forms work and perform on a more systematic basis.

What, then, is the bridge between urban morphology and urbanism, urban planning and design? Based on the foregoing, it can only be a bridge anyone chooses to build. There is no intrinsic, inescapable connection between the goal of an accurate description of the built environment and the goal of producing good urban form. To reflexively apply the principles brought to bear in the discussion above, building the bridge between urban morphology and urbanism becomes a matter of clearly identifying the problem, formulating the solutions and performance criteria and putting them into action. It also becomes a matter of persuading others that the goal is sufficiently important and the means is the best to achieve it.

In doing so, it is worth taking into account some of the variability and hazards in goal-directed behaviour. One point is that 'Desires become more specific in the pursuit of them. Desires are clarified by their means, and that can affect desires. For example, if people desire good lighting, it is not clear what type is meant by that: incandescent versus LED, overhead versus floor lamp, and so forth. Each could count as good lighting, but not all would turn out to be what is really desired. That becomes clearer in the process of figuring things out relative to the means available' (Liszka 2014, p. 466, quoting Peirce).

The hazard is that in contemplating the goal of providing good urban form at the most general level it might appear that urban morphology is the best means but once one gets into further detail, it becomes clear more is necessary. There is also variability in desires and goals. People seek or exercise freedom of latitude by applying a standard to different degrees at different times depending on their inclination. People may also seek the freedom to adjust their standards in response to different situations, which Peirce called the variable of longitude (Liszka 2014, p. 467).

Bearing this in mind, the question remains, what is the best means of building a bridge between urban morphology and urbanism, urban planning and design? One step in formulating an answer is to assess how urban morphology might contribute to a normative science of the built environment as sketched above. In principle the role of urban morphology would be limited to the planning and design stages of the process. The substages in the process would include: (i) a survey and assessment of the issues involved in the problem in order to refine the purpose; (ii) a search for possible solutions; (iii) mentally (or by models and simulations) working through the potential implications of the hypothesised solution and making adjustments as necessary (deduction); and (iv) comparing the proposed solution against established measures and/or other successful examples to check for anticipated performance and potential issues and making further adjustments as necessary (induction).

The sensing of the 'irritation' of a problem or problems and the identification of a purpose and the selection and putting forward of a solution as a means to resolve the problem (abduction or hypothesis) are necessarily judgements made by the individuals or groups—the various agents—involved in the process.

A further step in assessing the potential contribution of urban morphology to a normative science of the built environment is to be clear about what urban morphology has to offer. In general, the 'output' of urban morphology is a rigorous and accurate description of the built environment. As put forward elsewhere (Kropf and Malfroy 2013), the actual and in-principle subfields in urban morphology include the study and explication of: (i) the generic aspects and elements of form and their specific properties, (ii) generic and specific structure and relationships, (iii) regularities of development, (iv) evolution and diversification of form, (v) socio-physical performance and (vi) meaning.

For the most part, the descriptions are set in terms of types, processes/mechanisms, measures and dynamic models, supported by explanations (including original purposes) and interpretations.

5.6 Language, Teaching and Intuition

The next step would be to explore the combination of stages and potential contributions but for the immediate purpose of this chapter, it is well to bear in mind that the initial goal is to persuade a wider population that the 'problem' of building the bridge is a good end to pursue. The best means to achieve that goal may not be to use the more analytical, specialised language of accurate description/explanation and work on a more analogical basis. This effort should avoid deliberately or inadvertently disguising the normativity and implying that a description is a justification for a normative choice. Rather, it should be to ensure that building the bridge is an attractive proposition. To turn back, briefly, to Liszka and Peirce, 'Peirce provides a sketch of what formal properties an end or norm (an ideal of conduct) would have to have in order to be recommended as something worthy of pursuit. First, people would have to have a pro-attitude toward it. It must be an attractor. As Peirce describes it, the end must be admirable. Second, an ideal of conduct or an end worthy of pursuit is one that is consistent with other ends or ideals considered worthy of pursuit' (Liszka 2014, p. 473).

That is, the contributions of urban morphology to a normative science or, more immediately, to planning and urban design practitioners, might be more attractive as a means of meeting their needs and preferences when translated into non-specialist language that fits in with their own language and ideas. This might be criticised as 'dressing up' the contributions of urban morphology but it remains one of a number of options if the raw, untranslated language means the contributions remain unattractive. Another, or additional, option is to translate urban morphological principles by way of analogous ideas as expressed by other, non-morphologists. The result can be both to change the character of the language and more directly tie the principles to other ideas and purposes.

Some examples are as follows. Urban form is a material that is the medium of planning and urban design (Richard Sennett, *The Craftsman* 2005). Urban form is an evolving technology (Brian Arthur, *The nature of technology* 2011). The built environment is a library of ideas and design resource (the general idea of cultural evolution). Urban form is a living pattern of activities and interactions on which we operate (Manfredo Tafuri *Theories and history of architecture* 1976, quoting Walter Benjamin).

All of these openly sit within a normative realm of purposeful creation or generation of form. What I would highlight as a focal common point between them is the injunction or advice to designers: 'know your materials'. This is a core message in Sennett's book. If one is to become an adept and highly skilled designer, artist, craftsman or creator of any kind, it is essential to develop an intimate knowledge, understanding and practical familiarity with the medium in which you work. If you are going to make policy, set out guidance or generate designs that achieve their purposes and are successful, you need to know how cities go together and how they work—or don't.

And like other forms of art, craft and design, deep understanding often only comes with experience and channelling intuitive responses to the materials. That experience in turn often comes from experimentation, play and making mistakes. Because these are generally frowned upon or not possible with actual projects in the built environment (for obvious reasons), the channelling of experience in the form of active exploration and looking at places by walking around, site and context analysis, design exercises, theoretical exploration and modelling become all the more important.

Urban morphology provides a scaffold that helps tie those experiences together into a more coherent whole than would result from a random sequence. The scaffold of urban morphological concepts and analytical methods facilitates learning from experience, makes the experiences easier to recall and perhaps most importantly, helps each subsequent experience reinforce the previous. That is to say, the scaffold facilitates systematic comparison, which is the core of a rigorous approach. Teaching urban morphology to planning, urban design, architecture and landscape architecture students is therefore the foundation of any bridge between them. It is by no means a new idea but bears frequent repetition (Oliveira 2018).

A central part of the scaffold or framework of urban morphological concepts is the typological process, which, as noted above, might also be a framework for a normative science of the built environment. As with any evolutionary process, there is an apparent paradox with the typological process when viewed from a particular angle. It is the old chicken–egg paradox. Which came first, the city or the idea of the city? In all such cases, the paradox only arises when we leave out the dimension of time. The 'answer' is, a different kind of city in an ongoing recursive process. And while each step in any evolutionary process is necessary, there is a germ to them all, fathomable or not and however it might be induced or driven, which is the step of replication. Humans reproduce and we reproduce our cities. In that reproduction, our ideas about the future city are necessarily built out of our previous experience of existing cities and we then 'choose to apply' a particular idea at a given time to build or transform a physical city.

The basis for that choice is key. I would argue that the starting point for any choice is our own qualitative judgements about places. Our sensory and cognitive equipment is set up to respond to places and make intuitive judgements. Part of the process of learning to design places is learning to make use of those intuitions by bringing them into more conscious form. That is, intuitive judgements can and should be used as a prompt to interrogate why a place triggers a positive or negative response and identify what might contribute to the response. The step towards design is then to abstract the pattern or configuration of elements, as a diagram. We can then justify the use of that pattern in a new design (a normative claim) on the basis of its positive performance—on the assumption that it is used for the same or similar purpose as the original. The interrogation of our response should extend beyond the immediate pattern or configuration and trace its roots out in the wider context and supporting systems to ensure that if the pattern is abstracted as a 'type' it has the correct conditions to support it when transplanted. None of this is new as an idea but can be found in numerous sources: John Dewey's magisterial *Art as experience* (2005), the work of Saverio Muratori and the *scuola muratoriana*, and in *A pattern language* by Christopher Alexander, Sara Ishikawa and Murray Silverstein with Max Jacobson, Ingrid Fiksdahl-King and Shlomo Angel—to name a few.

The principal point here is to suggest that the scaffold of urban morphological concepts can facilitate the process of interrogation and abstraction and so can play a supporting role in design decisions. Once prompted by a positive response to a place, the scaffold provides a ready language of aspects, elements and relationships to articulate the abstracted diagram and to recognise its relative position and the set of conditions and connections that support it as a positive place in use. The same framework helps in recognising places with similar conditions and connections that might support the 'transplantation' of the type and provides a means of working through the implications of fitting it into a new place. Provided the purpose, the conditions and connections and the internal configuration of the type-diagram are sufficiently similar, the positive physical-emotional response to the original provides the justification for the normative design choice. One way or the other, that choice is a hypothesis that needs testing against the performance of the type in fulfilling its purpose.

5.7 A Case Study: The Stratford-upon-Avon Canal Quarter

As an anecdotal example, the following is an account of an abstraction and application in my capacity as both teacher and practitioner.

The abstraction was prompted by my response to a particular spot within the Paddington Basin redevelopment site in London. I visited the site on a field trip with urban design students and we chose the site in part because it presents both good and bad attributes. Our aim on the field trip is to get the students to directly experience a range of examples, good and bad, and to measure and analyse them and apply what they've learned and experienced in their design proposals. The site was formerly an industrial area lying along a canal and railway line and while wandering along the canal on my first visit, I came across a space just where an urban motorway overpass crosses the canal. Despite the presence of the major overhead road infrastructure and attendant traffic noise, the spot had a positive, relaxed feel and was surprisingly active. Just at the point where the overpass crosses the canal towpath, there is a junction with a route that leads into one of the main public spaces in the redevelopment and sculpture and ping pong tables had been placed within the wide junction space. There is also a cafe that fronts on to the space (see Fig. 5.2). What struck me was how active and positive the space was despite being in a relatively peripheral location within the development.

I took photographs and later interrogated the space more fully by looking at current and historic maps. What became clear early on was that the canal towpath provided an easy pedestrian and cycle route from residential areas into a central area with employment, retail and restaurants, facilitated by the redevelopment. A diagram of the space is shown in Fig. 5.3. In verbal terms, the spot is a widened junction space that creates a node on a longer distance pedestrian route from periphery to centre with a link to a larger public space, reinforced by the overpass and parallel water route of the canal. On the opposite side of the canal there are historic warehouse buildings that add visual interest along with the sculpture and ping pong tables within the space.

My conclusion from the analysis was that the success of the space arose in large part from its relative position within the network of routes and spaces, the amount of movement along the routes, fed by the activities at their ends and along them. Additional factors included the activity of the cafe fronting the space (and in turn

Fig. 5.2 View of the nodal space on the edge of the Paddington Basin redevelopment

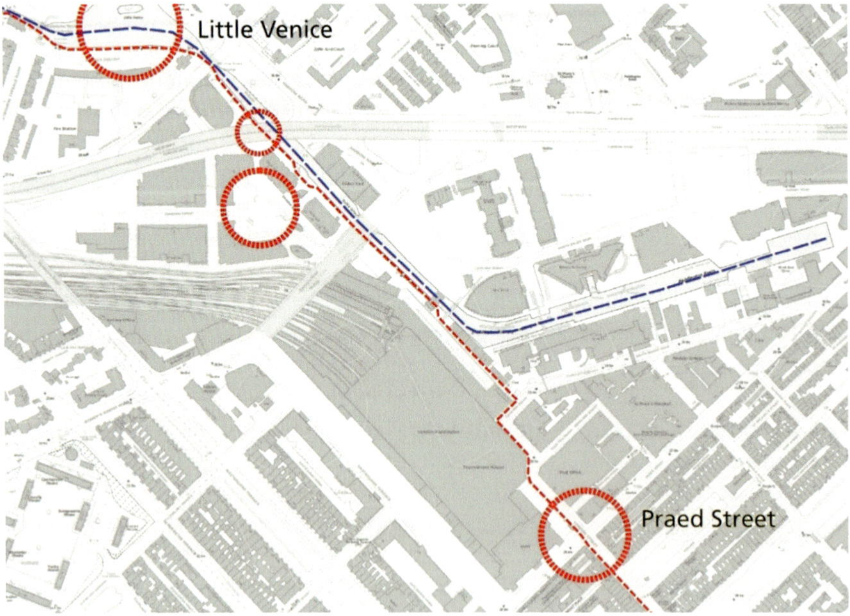

Fig. 5.3 Diagram of the nodal space in Paddington Basin, showing its relative position and 'roots' extending into the wider area

benefiting from the movement) as well as ping pong tables, the physical enclosure of the space, including the overpass, and the visual interest of the sculpture and historic buildings. One might test the relative contribution of the various factors in the performance of the space by imagining each factor in turn being taken away.

The opportunity to apply the type-diagram arose while working as a consultant on a redevelopment project in the Canal Quarter of Stratford-upon-Avon. The quarter is currently an industrial area, similarly located along a canal and railway. The area runs as a kind of 'fringe corridor' from just outside the boundary of the historic Medieval borough and the current edge of the settlement. The canal and towpath extend further into the historic core and connect to the main public space associated with the Royal Shakespeare Theatre. The towpath is therefore currently actively used by pedestrians and cyclists both for getting to work, school and shopping as well as for recreation. Because the redevelopment site is confined between the canal and railway, the principal access into the site is a road that crosses by bridge over the canal. At this point there is also a junction with the canal towpath with a connection to the main route running through the site. Just up from the junction is an open but private area currently used for parking.

The similarities between the two situations are by no means exact but, as illustrated in Fig. 5.4, the core common points are: a long-distance pedestrian/cycle route from the periphery to centre; a parallel water route (in both cases actively used for

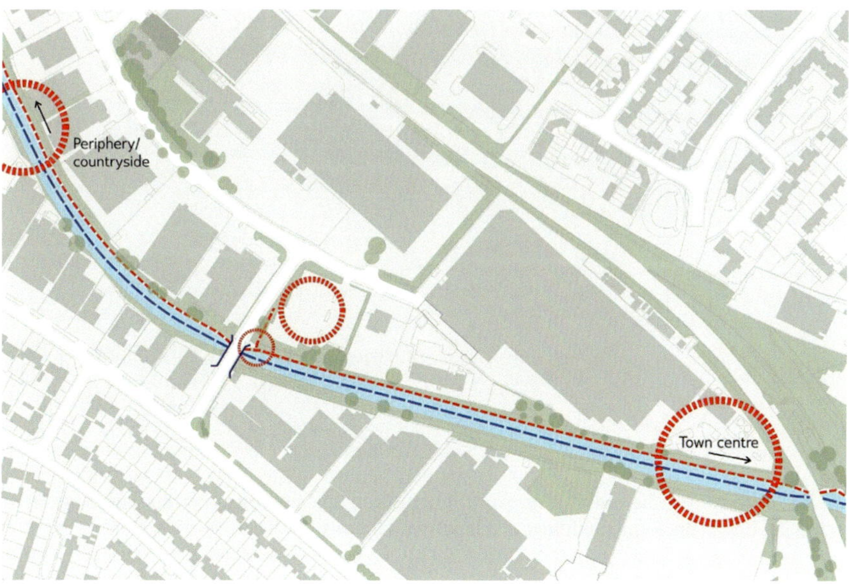

Fig. 5.4 Diagram of the nodal space along the canal in Stratford-upon-Avon within the Canal Quarter redevelopment area

recreation by canal boats); a junction with another route leading to another destination; a space at/within the junction; and other feature associated with and marking the position (overpass or bridge).

The similarity of basic elements and relationships prompted the idea that the location could serve the same purpose as the London example and so warranted exploring ways of elaborating the 'unformed' or incipient case on the Stratford site as part of the overall design framework for the redevelopment. The principal step in applying the type-diagram was to transform the car park into a public space connected to both the canal towpath and the main route crossing the canal and running through the site. A further step was to physically enclose the space with buildings with public fronts facing on to the public space, including a cafe near the canal (Fig. 5.5).

These design decisions constitute a normative claim about what ought to happen at this location as part of the redevelopment, justified by the positive performance of the abstracted type-diagram on which the design was based. The claim remains a hypothesis, as any design decision is, but tested to the extent of its similarity with the type-diagram. The hypothesis can only be fully tested by implementation and use.

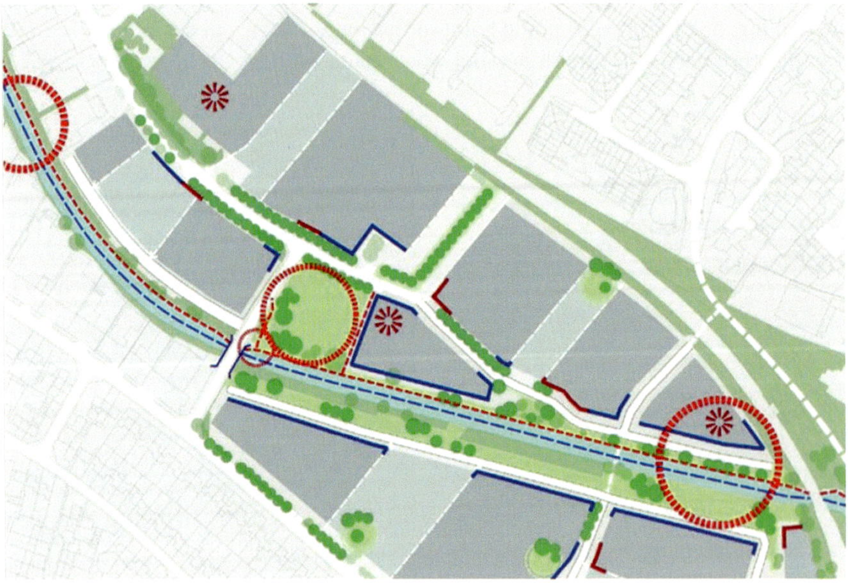

Fig. 5.5 Diagram of the initial framework design proposal, consolidating the nodal space by reinforcing it with enclosing buildings and landmark features

5.8 Conclusion

The argument being put forward and illustrated by the case study leads to what might seem an obvious point. Once incipient types get abstracted and repeated enough times because they perform their purpose well, they become codified types. Codified types are tested design solutions.

The reason large-scale house builders continue to produce standard house types is because they have proven successful. Similarly, the central core high-rise tower constructed with what Mark Jarzombek (2019) refers to as the *quadrivium* of concrete, steel, glass and plastic, is a global type that has shown itself to be eminently successful. The ubiquitous evidence of types, local and global, traditional and modern, across the planet suggests that the typological process is an accurate description of the process of producing and transforming the built environment. Given the similarity between the typological process and the hypothesised framework for a normative science of the built environment, it might be said that the framework is already in operation.

And as if we need reminding, crisis is part of the process. Even if the replication of types is warranted and justified by their performance, issues clearly remain. It becomes a question of who is making the decisions and for what purpose. Close replication of types is efficient, but who benefits from the efficiency? People are fond of invoking Jane Jacobs with the mantra of ‘the city is a problem of organised complexity’ but we end up forgetting why she became involved in the first place—the poor performance of the forms of development imposed on populations, often by public authorities. But then, thirty years before and thirty years later, in the 1930s and 1990s, the complaint was similar but against the poor performance of a different set of forms imposed mostly by private interests—the homogenisation of places by standardised suburban housing. Before that, from Friedrich Engels account of living conditions in Manchester to the work of George Peabody and later, municipal governments, the spectre of slum housing remained a persistent crisis for nearly a century (Tarn 1973).

A decade after Jane Jacobs published *The Death and Life of Great American Cities* (1961), Horst Rittel and Melvin Webber (1973) suggested that the kind of problem a city is, is a wicked problem. By all accounts it remains so. There are multiple, overlapping interests and purposes giving rise to multiple, overlapping problems that are difficult to define and have no definitive solution. In the end, the city really is an ethical problem. And while we might seek a normative science of ethics, in some ways the prospect of a ‘science’ skews our perspective to focus on the system as a whole or minute details. Our attention is drawn away from the central, human problems. To finish by quoting Peirce, ‘the only solid foundation for ethics lies in those facts of everyday life which no skeptical philosopher ever yet really called in question’ (Peirce 1965, cp 8.158, 1901).

A few of the most pressing at the moment are: income inequality, race and sex discrimination, resource depletion, pollution and climate change.

References

Alexander C, Ishikawa S, Silverstein M (1977) A pattern language. Oxford University Press, New York
Arthur WB (2011) The nature of technology: what it is and how it evolves. Free Press, New York
Batty M (2013) The new science of cities. MIT Press, Cambridge
Caniggia G, Maffei GL (2001) Architectural composition and building typology: interpreting basic buildings. Alinea, Firenze
Dewey J (2005) Art as experience. Perigee Books, New York
Hume D (1964) A treatise of human nature. Clarendon Press, Oxford
Jacobs J (1961) The death and life of great American cities. Random House, New York
Jarzombek M (2019) The Quadrivium industrial complex. e-flux. https://www.e-flux.com/architecture/overgrowth/296508/the-quadrivium-industrial-complex/. Accessed 7 Aug 2020
Kropf K (2001) (ed) Stratford-on-Avon District Design Guide. Stratford-on-Avon District Council, Stratford-upon-Avon
Kropf K (2014) Agents and agency, learning and emergence in the built environment: a theoretical excursion. In: Larkham P, Conzen M (eds) Shapers of urban form. Routledge, London, pp 303–322
Kropf K (2017) Handbook of Urban Morphology. Wiley, Chichester
Kropf K, Malfroy S (2013) What is urban morphology supposed to be about: specialization and the growth of a discipline. Urban Morphol 17:128–131
Liszka JJ (2014) Peirce's idea of ethics as a normative science. Trans Charles S. Peirce Soc 50:459–479
Mehaffy M (2014a) What can a 'science of cities' offer planners? Planetizen—urban planning news, jobs, and education. https://www.planetizen.com/node/69957. Accessed 7 Aug 2020
Mehaffy M (2014b) Lessons from an emerging 'science of cities'. Meeting of the minds. https://meetingoftheminds.org/lessons-emerging-science-cities-10828. Accessed 7 Aug 2020
Oliveira V (2016) Urban morphology: an introduction to the study of the physical form of cities. Springer, Cham
Oliveira V (ed) (2018) Teaching urban morphology. Springer, Cham
Oliveira V, Silva M, Samuels I (2014) Urban morphological research and planning practice: a Portuguese assessment. Urban Morphol 18:23–39
Peirce CS (1965) Collected papers of Charles Sanders Peirce. Harvard University Press, Cambridge
Peirce CS (1983) A brief intellectual autobiography by Charles Sanders Peirce. Am J Semiot 2:61–83
Petruccioli A (2007) After amnesia: learning from the Islamic Mediterranean urban fabric. Dipartimento di Ingegneria Civile e Architettura Politecnico di Bari, Bari
Putnam H (2002) The collapse of the fact/value dichotomy and other essays. Harvard University Press, Cambridge
Rittel H, Webber M (1973) Dilemmas in a general theory of planning. Policy Sc 4:155–169
Samuels I, Pattacini L (1997) From description to prescription: reflections on the use of a morphological approach in design guidance. Urban Des Int 2:81–91
Sennett R (2005) The craftsman. Yale University Press, New Haven
Tafuri M (1976) Theories and history of architecture. Harper & Row, New York
Tarn JN (1973) Five per cent philanthropy: an account of housing in urban areas between 1840 and 1914. Cambridge University Press, Cambridge
Wilson A (2012) The science of cities and regions: lectures on mathematical model design. Springer, Cham, Netherlands. https://doi.org/10.1007/978-94-007-2266-8

Part III
Urban Morphology and Architecture

Chapter 6
Morphology and Typology: A Village as a Cultural and Environmental Process

Wowo Ding

Abstract A village is a rural settlement that has grown in close relation with the environment and geography, being constantly shaped by the social and cultural development. The most powerful force fostering environmental change in China is the undergoing urbanization, which has changed, not only the form of cities but also, the form of rural settlements, including villages. In general, most villages have disappeared during the process of urban growth. However, some remain, by various reasons, although under very controversial conditions, both culturally and environmentally. The project for the small community of Huangzhuang in the northern part of China's Jiangsu province is examined in this chapter as a practice case in village revitalization. Design efforts were made to integrate contemporary facilities in the existing settlement fabric. The architectural interventions, though modest, have catalyzed a new-found confidence among residents concerning their communal assets. These interventions have revalued everyday resources of people and place as the basis for making a lifestyle in decline more liveable. The chapter will show how design can be based on research and conclude on the meaning of using morphological research in practice, debating why practice needs research, how it works and in which ways it can be done.

Keyword Rural settlement · Village morphology · Housing typology · Design by research

6.1 Introduction

China has been rapidly urbanized during the 40 years of reform and opening up to the exterior. In processes of urbanization, cities and villages are developed through mutual influences. Urbanization brings both the agglomeration of urban population and the expansion of urban scales, and the sharp decline of rural population and the disappearance of villages. In urbanization processes, villages have changed due to

W. Ding (✉)
Nanjing University, Nanjing, China
e-mail: dww@nju.edu.cn

V. Oliveira (ed.), *Morphological Research in Planning, Urban Design and Architecture*, The Urban Book Series, https://doi.org/10.1007/978-3-030-66460-2_6

the continuous acceptance of urban civilization, which is not only a living behaviour, but also affects the village forms and geographical landscapes. Yet, the change of village forms in the process of urbanization has not received enough attention from research. Indeed, research on villages has been mostly interested in how these have been formed in the past.

China has experienced rapid urbanization. Although urban land is still owned by the State, the right to use it is obtained through market transactions. Therefore, in a capital-driven's urban development, the evolution of urban form still conforms to usual processes of transformation. In the countryside, the state-owned land system and the long-term dual system of urban and rural household registration make the evolution of rural settlements' forms in the context of urbanization processes a special phenomenon. Understanding rural settlements in urbanization processes shall not be seen simply as rural affairs—with the entire social, cultural, and economic actions in the built forms of urbanization.

In some cases, with the expansion of cities, rural settlements have not declined, but grown. New rural settlements have formed, involving houses of a uniform type and style, or residential areas composed of city-type apartments located in the countryside—the so-called new villages of farmers. The emergence of this phenomenon completely changed the rural landscape and the existing interpretation framework on the formation of villages.

In recent years, as the urbanization processes have entered a new stage of development, the construction of new urban forms has increased and the quality of physical space has started to improve. The latter lead to the inclusion of the quality of rural environment in the agenda. The terms 'tradition' and 'locality' have become associated with the quality of villages. However, as a large part of the rural population has moved into cities, the countryside started to change and many rural houses become damaged or vacant. Rural settlements labeled as 'new villages' are no longer the 'villages' that we knew, with a notable history, culture and locality, as they are now planned and designed as modern urban settlements.

Against this background, a Rural Revitalization Strategy, as part of the political agenda, has been proposed, aiming at rectifying a number of urban–rural disparities created by decades of uneven development. The Strategy suggests the revival of fringe economies through land use modification, infrastructure improvement and village renewal. Many pilot projects promoting rural tourism are already underway as a part of a comprehensive effort to raise local livelihoods' quality. As part of this process, many planners and architects were encouraged to go to rural areas for new planning and designing exercises, in an attempt to restore the 'traditional forms' of villages, based on the reading of inherited forms. The author of this chapter, an architect, was also invited to be involved in this work.

6.2 The Case Study of Huangzhuang

Huangzhuang is located in the Gaobao Lakes Plain, north of the Yangtze River and east of the Beijing-Hangzhou Grand Canal in Jiangsu province. Its west side is the artificially excavated Huaihe River Water Channel, making the whole area a peninsula, extending to the Gaoyou Lakes. Huangzhuang Village is 1.8 km from Taji, a small town in the peninsula (Fig. 6.1).

Huangzhuang has the typical form of a natural village of the region—the strip-shaped villages (Fig. 6.2). This kind of village is called Zhuangtai, a village built on a high platform. In the region, all villages are formed by two rows of houses with a straight river in the middle. The front and back of the two rows are farmlands. This regular arrangement constitutes a particular structure of the region, and the village form is tightly related with the land form (Fig. 6.3).

Unlike ordinary villages, strip villages do not have entrance *plazas*, which are usually important public spaces. Therefore, there are no public spaces for residents to gather. Planners and architects often conceive public spaces in their design proposals, but these are often intrusive to the existing village forms.

The Huangzhuang village has 168 households and 757 residents. Only one-third of the houses are occupied all year; one-third are temporary residences and one-third of the houses are not occupied. The farmers' houses are built according to a simple type, with three or four rooms connected by a corridor, forming the main body of the house. Most of the houses have only one floor. There is no bathroom or kitchen in the main building. The kitchen is a small hut in the south side of the house, while the toilet is just a shed placed behind the house (Fig. 6.4).

The quality of the houses is generally good. Yet, those that have not been inhabited for many years are beginning to decay due to the lack of maintenance. Although these houses are uninhabited, they have not been abandoned. Their owners work and live in cities. As these unoccupied houses are scattered throughout the strip villages, they have a negative impact on the surrounding households and further weaken the public space. How to deal with these uninhabited houses and how to evaluate their value are difficult issues for planners and architects.

Fig. 6.1 Maps of China and Jiangsu province, Gaobao Lakes Plain, and Taji town

Fig. 6.2 Aerial photo of Huangzhuang village—typical linear village in Gaobao Lakes Plain (*Source* Project team lead by Ding)

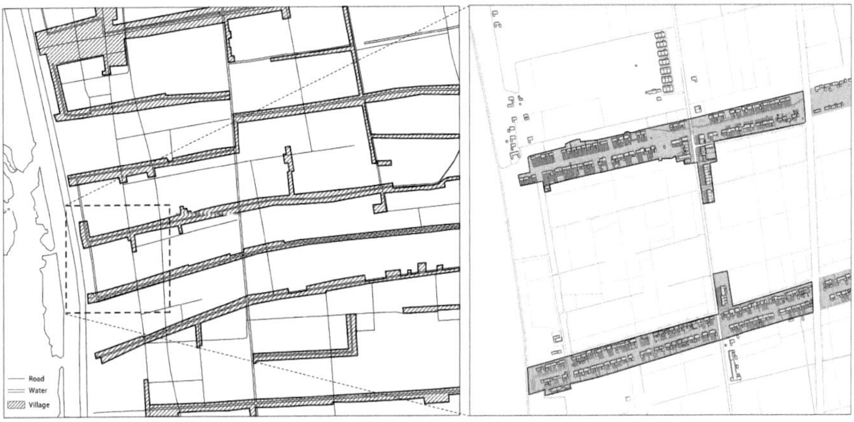

Fig. 6.3 The strip villages of the region (left) and Huangzhuang village in two strips (right)

A village is a form of aggregation, of living in a farming society. It reflects the way of working and living, representing a comprehensive mapping of the relationship between human and nature, land and production, and also between different human beings. The village form is the result of evolution through time. As such, a revitalization project of this nature it is not just a design work; its significance lies in recovering the rural environment as 'it was originally'. The design process should be based on research. There are three factors to be investigated and studied, as follows.

The first is the study of geography and environment. Villages are intimately related to the land. Settlement geography is essentially concerned with the processes of land settlements and the forms of these settlements. The purpose is to know the origin of genuine agglomerated forms and its transformation forces.

The second factor is the investigation of social development, including policy analysis. The aim is to know what is the role of villages in processes of urbanization, how different policies act in the formation and transformation of villages and what are the fundamental directions for the future.

The third aspect is the study on rural settlement typology. Building types are related to the way of living. The purpose is to understand what are the current requirements for rural settlements, why are the traditional types of farmhouses being abandoned (or which parts are being abandoned), how to develop a new building type and what is the role of design in this process.

As villages are the structural frameworks for agrarian settlements in many regions of the world, it is important for our general knowledge of societal development to establish under what circumstances, and in what way, village formation occurs in various areas (Callmer 1991). Morphological research on villages do not only seek to explain the formation processes but also to define the bases for design.

Fig. 6.4 Plan of the basic type of farm house and views of the houses (*Source* photographs by Wowo Ding)

6.3 Settlement Geography

Settlement geography developed in the 1960s is essentially concerned with the processes of land settlements and the forms of these settlements. The spatial distributions under analysis are usually those of dwelling places (Stone 1965). Settlement forms are the spatial realization of transformations in population. There are four sources of location theory: central place theory, diffusion theory, ecological distribution theory and the fourth category of generalizations encompassing aspects of all these three, morphological laws (Hudson 1969), which are the basis of this work.

The topographical conditions and landscapes, together with the social and economic aspects, are of crucial importance for a profound understanding of the transformation of settlement up to the relatively stable pattern of historical times (Callmer 1991). The morphological composition and evolution of rural settlements involve three factors. The first is topographical conditions. The landforms are directly related to the products of agricultural production and farming methods, and thus to the size of the population that can be supported and to the size of its settlements.

Chinese scholars have considered the importance of landforms in terms of the conditions and the frameworks for dynamic spatial changes and family composition with residential culture. Rural settlements have been formed over a long period in history and different geographical contexts in China. The density and distribution of different natural conditions, socio-economic conditions, historical development and living habits have obvious differences and are the product of combination of various factors (Liu et al. 2010). Most of the existing rural settlements in China are developed based on traditional settlements. Their distribution and scale are in line with the social and economic conditions of their time. Productivity and the ability of people to control nature at that particular time are in harmony with the production and environment (Jin 1988). The plains of Northern Jiangsu have a high number of rivers and lakes, and the form of the water network is the main framework for the landforms. There are substantial differences between villages settled by lakes and by rivers. Not only the shapes of the water networks affect the patterns of villages, but also their period of formation determines the forms of villages (Ding 2001).

Gaoyou, where Huangzhuang is located, is one of the many lakes in the area called Gaobao Lakes—a group of lakes west of the Grand Canal, in the central part of Jiangsu Province. Chinese geographers have recently carried out an important research in this area. They have concluded that, under the influence of the changing relationship between the Yellow and Huaihe rivers and land reclamation activities, the shapes of the Gaobao Lakes have changed dramatically in the last 300 years (Yang and Han 2018). According to this research, the area where Huangzhuang is located in a new land formed by the shrinkage of the Gaobao Lakes, formed before 1868 (Fig. 6.5).

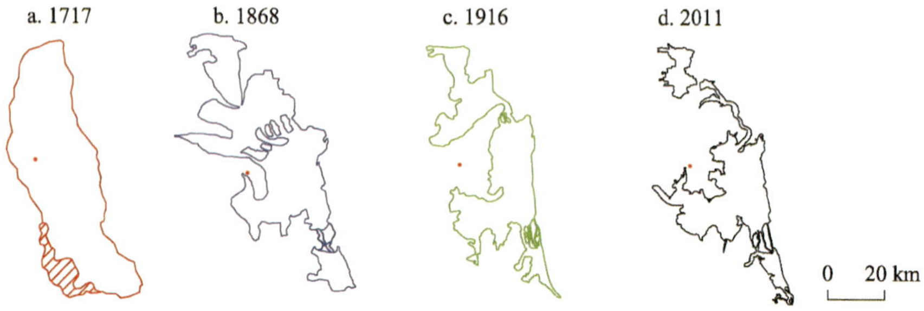

Fig. 6.5 The reconstructed shape of the Gaobao lakes based on surveyed maps, 1717–2011 (*Source* Yang and Han 2018)

A number of ancient maps of the area have been collected and studied. The comparison of these maps supports the argument on the formation time of the area where Huangzhuang is located. The most important evidence is a map from the twenty-first year of the Guangxu Period of the Qing Dynasty (1895)—Taji town had already been marked on the map (Zhu 1895) where Huangzhuang belonged. Prior to that, in a map drawn in the Xianfeng Period (1807–1809), Taji town was still located in the Gaoyou Lake (Fig. 6.6).

One of the reasons for the formation of this area is the entrance of sediments from the Yellow River into the Gaobao Lakes, through the water of the Hongze Lake, leading to the silting and shrinking of the lakes. Another reason is that the polder fields have become an important activity of agricultural production at that time, and the reclaiming land from the lakes was the best way to expand cultivated land, as the massively growing population needed more land to produce more food. The land where Huangzhuang is located is the product of polder farming, and the artificially filled up land is the product of lakes enclosing. (Liao 1992; Yang and Han 2018).

The process of filling up fields is a long one, with the expansion of farming fields, piece by piece, gradually advancing into the lakes. In the new land, artificial canals both drain and irrigate the different plots. Along both sides of the canals, houses are built on higher grounds. The canals between the two rows of houses provide water to the inhabitants. Water became the centre of their daily lives.

The morphological analysis clearly explains the mechanism of formation of strip villages and makes evident the important role of water channels in farming and daily lives.

Fig. 6.6 Map of waterways in northern Jiangsu Province, 1807–1809—top, and Gaoyou Lake in 1895—bottom (*Source* public domain)

6.4 Land Policies and Urbanization

Another influence on the change of rural settlements and village patterns is land policy and the household registration policy. Since the 1950s, China has established a dual system of urban and rural areas, called the dualistic structure system of urban and rural areas. The most prominent point of the urban–rural dual system is that the user registration system divides people into two categories: urban people and rural people. The State implements different distribution methods of benefits for different types of population. Essentially, it limits the mutual flow of population by establishing the attributes of household registration. During the planned economy era, the urban–rural dual system was beneficial to the State management, but after the reform and opening up, especially in the processes of urbanization, the drawbacks of the urban–rural dual system have been particularly relevant.

In rural areas, a resident has the right to apply for a homestead from the Village Community to which he belongs based on the principle of one homestead per family. Usually, when a man from a family grows up and gets married to start a new family, he can apply for the land. The homestead is a distribution system. For the sake of fairness, the size of the homestead is determined by the Village Collective according to the amount of land ownership. Because of the limited household registration, it is difficult for the population to move around, so the density of strip villages is higher as population grows—the size of each household is almost the same.

The growth of cities in recent years has greatly increased the demand for immigrants, and many rural people have migrated to large cities to find a job. In order to stabilize the migrant population, urban policies have become increasingly tolerant, offering rural population the possibility of converting to the types of urban household registration. In addition, for those without an urban household registration, cities have provided various facilities, enabling the permanence of rural population. On the other hand, the advantages of free access to residential plots are becoming apparent, and many rural people who have been living in cities for a long time are reluctant to give up their status in order to keep the houses they own in their villages.

This is the main reason why the form of villages is not shrinking due to the loss of population. There are very few people who live in villages all year, but the unoccupied houses are, in fact, occupied. This phenomenon is called the hollowing of villages.

Another phenomenon is the impact of changes in rural municipal facilities on the entire water environment. Along with urbanization, rural areas have begun to improve their standards of living to urban standards, mainly through infrastructure development. People in the countryside began to abandon the use of water from rivers or lakes in favour of clean tap water, and in the case of Huangzhuang, the canal between the two rows of houses was gradually abandoned. On the other hand, wealthy residents acquired agricultural machinery and family cars—the village roads were upgraded from pedestrian to carriageway. With the introduction of the carriageway, the transport functions of the canal disappeared and, eventually, the canal has changed from the centre to the backyard.

6.5 Housing Types

Most sociologists have argued that two social relationships are crucial for the establishment of natural villages: blood relationships (the most important) and geographical relationships. As the Chinese sociologist Fei Xiaotong has stated 'Blood ties are a stabilizing force. In a stable society, geographical ties are only a projection of blood ties, and a reproduction of the population from one generation to the next' (Fei 2001). Huangzhuang is a village with the Huang Family as its main constituent, and all other surnames have gradually migrated over since. In modern societies, the concept of family is fading, and with the allocation of residential land, the family imprint is hardly reflected in the village forms. Thus, the house type became an important vehicle for reflecting the family lineage (Fig. 6.7).

As above mentioned, the house type in Huangzhuang is the same as in most farmhouses in the area: it consists of three rooms lined up in a row, with a southern corridor connecting the three rooms—maintaining the basic prototype of the traditional Chinese house. Although the main material of the current house is bricks and some concrete, this type of house still follows the rules of the wooden structure.

In Chinese traditional housing typology, the 3-bay house—three rooms in a row—is the basic type (Ding 2001). The middle bay, called *tangwu* (hall), is the most important. This bay is at the centre of the family and it has three main functions. Firstly, it is the place where the ancestors of the family are worshipped, and where the tablets (a record of honour) of the family ancestors are placed. Secondly, the *tangwu* is a place to receive guests or develop family business. And thirdly, the *tangwu* is also the family's dining room, where family members have dinner together. The rooms on both sides of the house are the bedrooms, and are generally occupied by two different generations. When the younger generations get married, they apply, as adults, for new houses on other sites. When the composition of family members is more complex, there are also a small number of four-room types. Regardless of the types, the *tangwu* is always centred, with the same furnishing, where the ancestry and heritage of the family can be shown.

There are neither kitchens nor toilets in 3-bay rooms. The kitchen is a secondary room and it is built in the south corner. There are basically no toilets. The small shed behind the house is used during the day and the close-stool is used at night. This type continues until today, with a few changes.

A few new two-layer houses have been built in the village. The ground floor maintains the traditional layout of 3-bay, with the *tangwu* in the centre of the house and the other rooms used for stacking, sundries or bedrooms for the elderly. The first floor is usually a bedroom. In this new type, the addition of a bathroom improves the comfort of living, but it is not linked to a sewer system. Sewage simply drains to the canal, behind of, or in front of, the house increasing the pollution of the water networks.

Although two-thirds of the houses in the village are usually uninhabited, no matter where people live or work, according to the tradition, they return to their village homes during the Chinese New Year. Ancestor worship is the main event, followed

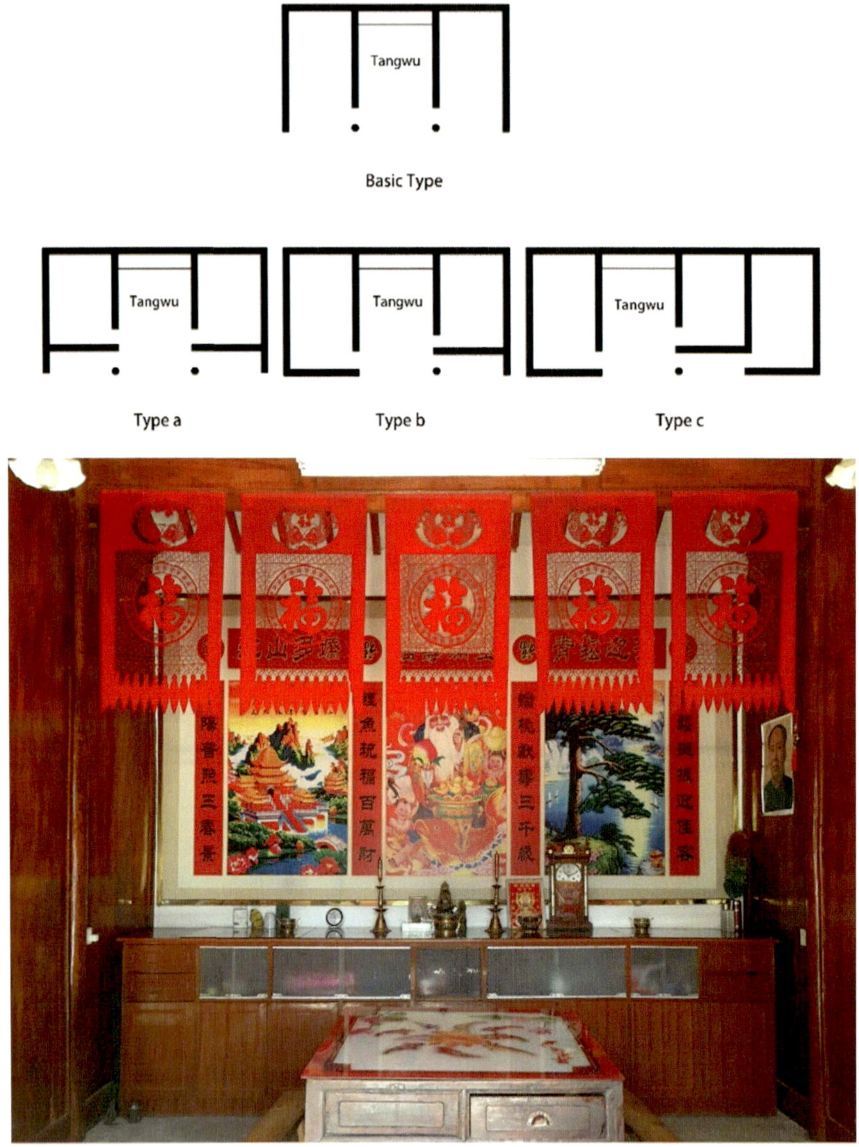

Fig. 6.7 Basic type and its variations (top) and interior view of the *tangwu* (*Source* Project team lead by Ding)

by relatives visiting each other. At that time of the year, the village is very busy and almost every house is occupied.

According to research, the acceptance of existing houses by residents varies considerably with age. The main problem identified is the lack of modern sanitation facilities. Generally speaking, the elderly are very accustomed to use this house in the old way; adults do not like it, but they can use it; and young people and children who have lived in cities for a long time do not accept the house without a bathroom—they are willing to come back to ancestors and gatherings, but do not want to live in the old house in the village. The result is that young couples and children returning to the village for the Chinese New Year stay in hotels in the town, during the night, and then return to their homes in the village, during the day. Therefore, the functions of traditional houses face major challenges.

6.6 Site Plan

The study of geographical changes in the region and of the causes of village formation have led us to realize that Huangzhuang is located in a new land in an ancient region. This was formed mainly as a result of the expansion of agricultural production; the formation of striped villages is later (it explains the genesis of the village morphological features). Furthermore, we are aware of the east-west strip villages on the land gained to the Gaoyou Lakes. Finally, the form of the two rows of houses bordering the canal reflects the relationship between people and place. Therefore, the careful preservation of the morphogenetic imprint of the village is the basis for the village planning and design.

Another aspect that needs to be considered is the role of villages in the processes of urbanization and the significance of its existence. What are the aspirations of the village residents? And what are the demands of villagers who live and work away from their villages?

Field investigation shows that there is still a considerable number of people who are living in the countryside: some prefer to live in their own villages closely to the town where they have their jobs; a large number of old people and children are living in villages, as old people like rural environment and the way of living, and children get an easier life being with their grandparents than with their parents in cities, where the latter have busy lives. Those who work in cities are reluctant to give up their houses in villages, not only because of the quality of the house but also because it represents a link to the family tradition. In homecoming in Chinese New Year and in other festivals it means 'nostalgia' for most Chinese.

The recent affluence of different people, with a variety of needs, to the countryside has motivated research on the transformation of rural areas. There are four types of people, and purposes for, living in, or going to, the countryside: the most popular is the purpose of weekend vacation, where villages around cities receive people from these cities in the weekends; the second purpose is countryside vacation, where people choose a more comfortable environment to rest for a week or even a month; third, people living in cities choose to go to villages for retirement; and, finally, the last type includes people returning to their hometowns to start their own businesses and agricultural technicians who work in the countryside and choose to live in villages.

Based on our research, we have established an implementation plan for the renewal of villages. We believe that the form of villages records the genesis and history of their growth and that its basic structure should be preserved. In order to respond to the new needs and diverse population, the renewal design shall increase infrastructure systems, build micro sewerage facilities and connect every household.

We propose to dredge the canal between two rows of houses and to clean up its banks, allowing the river to become once again at the centre of the residents' lives. Along the embankment of the river, a new cycling track was built by the government to pass by the village, connecting it along the river loops. We added a motor vehicle road and a walking path into the village. Finally, a pavilion was built at the entrance to the village (its position being benefited by relief). The pavilion serves two purposes, to identify the village from the road along the river and to provide a full view of the village (Fig. 6.8).

6.7 Building Type and Design

As above mentioned, Huangzhuang has a relatively unified type of house, the 3-bay building and a small kitchen as auxiliary building. When a resident applies for the construction of a new building, the Village Committee provides a site with a proper shape to fit the specific house. This means that the extant plot structure has a strong relationship with the traditional type of building. It also means that the size and shape of the plot, the type of house and the form of the village are deeply interrelated. Due to typological roots in traditional culture and local habits (Quincy 1788), the basic type of building in the village does not only reflect the past way of living but also provides the physical bases and references for design.

We have found that the *tangwu*, in the 3-bay house, is the heart of the family spirit and the centre of the annual Chinese New Year ancestor worship. Indeed, the *tangwu* is a key part of the unchangeable 3-bay room structure. On the other hand, without sanitation facilities, the traditional farmhouse can no longer meet contemporary needs of adults, youth and children. Even when they return home to pay respect to their ancestors, during the Chinese New Year, they do not want to stay at home, and look for a small hotel. We understand that we must transform the building from a traditional to a contemporary house, while preserving the structural and basic elements.

At the beginning of the design process, the mayor, on behalf of the villagers, expressed the hope that our project would increase the value of the vacant homes. The idea was that the Village Committee would manage the renovated houses on behalf of the owners and would have the right to rent them, offering a share of the profits to the owners. The Village Committee contacted all owners of the vacant

Fig. 6.8 Project: site plan—with maintenance of the original form features, entrance pavilion and three renovated houses (*Source* Project team lead by Ding)

buildings and none wanted to sell their homes. All wanted to rent the houses and have agree with the renovation. The additional condition was to keep the main building type—and the *tangwu*—unchanged. Initially, only four families were willing to rent the houses to the Village Committee. The project started with these four houses.

Based on the four houses, we started the design experiment. In order to meet the needs of various groups of people for short-term or long-term housing, while maintaining the same basic type, we propose to add bathrooms, reorganize the hall to accommodate the new additions, increase the connection between the kitchen and the main buildings, and organize the outdoor spaces.

The first task was adding bathrooms to these four houses. The selected option was to locate the bathrooms outside the main buildings, preserving the basic types of the original houses—in their back or side-by-side.

The *tangwu* was kept in each house, redesigning the space as a sitting room for guests, maintaining the long table waiting for family members to return in the Chinese New Year.

In the original houses, there was no connection between the kitchen and the main building. Food was taken from the kitchen to the dining room through the exterior area of the plot. This was unacceptable for today standards of living. To improve the comfort of the original houses, we have designed a simple wooden corridor between the main building and the kitchen. The corridor provides shelter from rain and ventilation, as well as an outdoor seating area, which is suitable for guests who come to the countryside for leisure.

The project included the reorganization of the exterior space in front of, and behind, the house. Villages in this area have exterior spaces in front of their houses, for drying or temporary storage in the harvest season. Nowadays, most of the village residents are elderly and children, who are not engaged in agriculture. As such, there is no need for drying the crops in front of the houses (where drying sites have turned into junkyards). However, such sites are precious for guests, coming from the city to relax—accordingly we have changed the original drying places into gardens and courtyards (Fig. 6.9).

The purpose of our project was to demonstrate the village residents how an existing house could be adapted to modern life through simple modifications. The renovated house can meet the needs of young people and also receive guests from cities. In fact, strategically thinking, rural housing in the processes of urbanization must be adapted from single to multiple functions. Furthermore, this renovation process offers important insights on the renewal of building types (Figs. 6.10, 6.11, 6.12, 6.13).

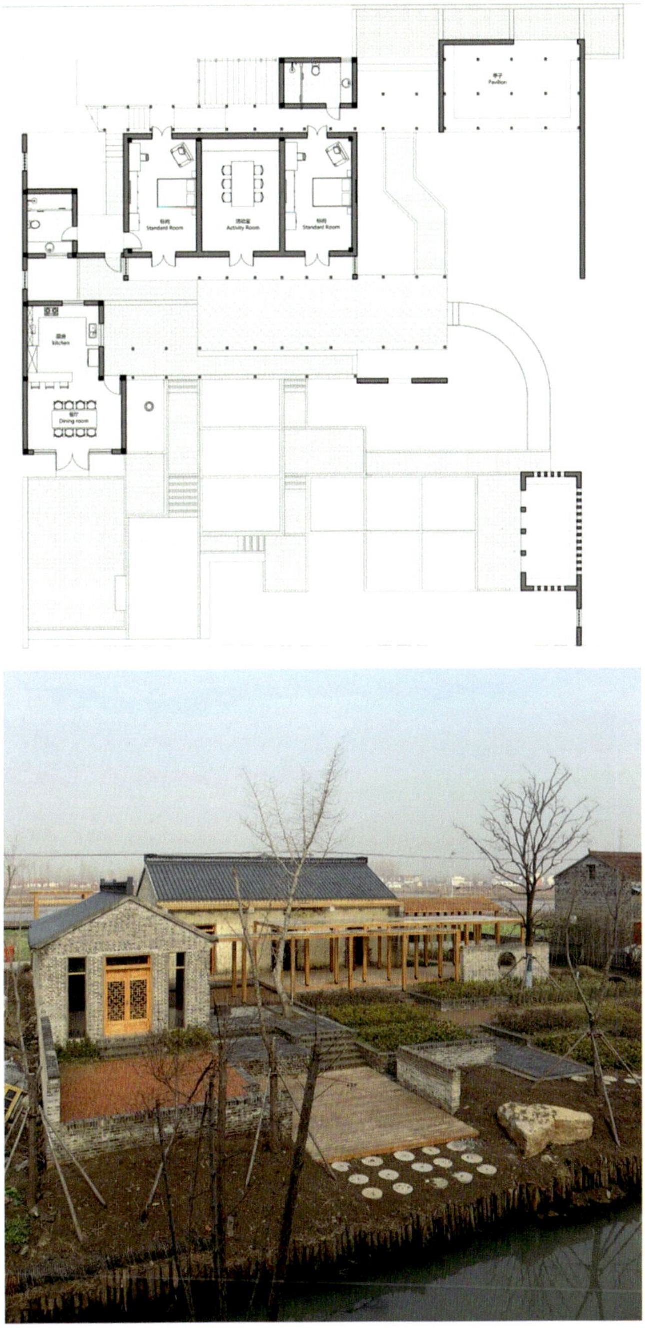

Fig. 6.9 Project: plan and view of one of the houses (*Source* photograph by Wowo Ding)

Fig. 6.10 Renovation of sanitary facilities, before (bottom) and after (top) (*Source* Project team lead by Ding)

Fig. 6.11 Maintaining the hall and updating facilities, before (bottom) and after (top) (*Source* photographs by Wowo Ding)

Fig. 6.12 Connecting the main building with the kitchen, before (bottom) and after (top) (*Source* Project team lead by Ding)

Fig. 6.13 Cleaning the open space in front of the house for gardening, before (bottom) and after (top) (*Source* Project team lead by Ding)

6.8 Conclusions

The Huangzhuang project started in 2017 and has been completed in 2019. For its authors, this is not only a village renewal project, but also the realization of morphological and typological research in practice. Morphological research in practice is mainly manifested in two aspects: the study of landform change and social transformation. Research of geomorphological change made evident three crucial aspects: Huangzhuang is located in a region that was formed relatively late, its formation processes record the entire process of erosion of lakes by artificial farming, and it contains the elements to understand the morphogenesis of strip villages. The study of social change made evident that housing and land policies have played a key role in the expansion of village forms. The urban and rural dual system is also important to understand why the village has not gradually disappeared but hollow out in the process of urbanization. Urban morphology helps to describe and explain the object and to design strategies and actions. Typological research offers guidance for the understanding and analysis of traditional types, rather than simply collecting the formal language of local buildings. Together with urban morphology, it lays the foundation for design.

After the rural revitalization project was completed, the relationship between the farmland, the canal and the village remained the same, and the fundamental characteristics of the local strip village have been maintained. The canal has, once again, become the centre of residents' daily lives and a popular recreational area for both villagers and visitors. Four pilot houses were fully approved by the owners and generally acknowledged by villagers. As a result, more villagers are willing to hand over their houses to the Village Committee for renovation and management. Furthermore, villagers have also begun to clean up the sites around their houses, changing their grounds into small gardens.

Acknowledgements The members of the project team are Wowo Ding, Wei You, Lian Tang and Qian Li. The author would like to thank the students involved in the project for their contributions.

References

Callmer J (1991) The Process of village formation. Ecol Bull 41:337–349

Ding W (2001) The generation of a village—the study of villages in the ZJG region. ETH University Press, Zurich

Fei X (2001) Jiangcun Economy. Commercial Press, Shanghai

Hudson JC (1969) A location theory for rural settlement. Ann Assoc Am Geogr 59:365–381

Jin Q (1988) The History and current trends of research on rural settlement geograpgy in China. Acta Geogr Sin 43:311–316

Liao G (1992) Formation and evolution of the Gaoyou Lake, Jiangsu province. Acta Geogr Sin 47:139–145

Liu YS, Liu Y, Chen YF, Long HL (2010) The process and driving forces of rural hollowing in China under rapid urbanization. J Geog Sci 20:876–888

Quincy Q (1788) Type. Encyclopedie methodique: Architecture. Panckoucke, Paris
Stone KH (1965) The development of a focus for the Geography of settlement. Eco Geogra 41:346–355
Yang X, Han Z (2018) The change of the Gaobao lakes and its driving forces (1717–2011). Acta Geogr Sin 73:129–137
Zhu K (1895) Map of Jiangsu Province

Chapter 7
The Terni Cemetery—Considerations on the Relationship Between Reading and Design

Giuseppe Strappa

Abstract This chapter deals with the theme of the relationship between the morphological reading of the built reality and the architectural design. The author uses the project and construction of the Terni Cemetery to propose an interpretation of the problem. The theme of the cemetery seems particularly congenial to the subject constituting a city analogous to the real one, containing the problems of morphological analysis, architectural language, and aesthetic synthesis. The architecture of cemeteries, moreover, is perhaps one of the field of our discipline in which the changes produced by industrial civilization caused the greatest losses. The text supports the thesis that the design of cemeteries must follow some general principles related to dyads of notions such as organism and congruence, enclosure and covering, hierarchization and specialization. One of the fundamental problems, from a morphological point of view, in the Terni cemetery design, was to identify the ways of transforming nature into architecture. The choice of using plastic-masonry structures was due to two categories of considerations. The first is linked to the notion of 'cultural area'. All buildings in the Umbrian area have strong stone features that constitute one of the most vital factors in the cultural continuity in this territory over centuries. The second is linked to the organic and 'didactic' character of this material, establishing the necessary continuity that regulates the position of elements: a good masonry building is linked to physical laws that impose an organic hierarchy between the parties. The chapter briefly addresses the question of architectural language. The author proposes to recognize in the term 'architectural language' the value of personal use of a common and shared language based on the 'materiality' of architecture, which should be understood, studied and respected.

Keyword Architectural reading · Architectural language · Design

---

G. Strappa (✉)
Sapienza Università Di Roma, Rome, Italy
e-mail: gstrappa@yahoo.com

V. Oliveira (ed.), *Morphological Research in Planning, Urban Design and Architecture*, The Urban Book Series, https://doi.org/10.1007/978-3-030-66460-2_7

7.1 Introduction

The problem of the link between morphological reading and design has always been complex and often contradictory. Vitruvius, who first raised the problem of a general theory of the built form, when erecting the Basilica of *Fanum Fortunae* did not follow the principles he himself had laid down. Perhaps he had wanted to write, not a manual based on the building reality of his time but, a treatise based on ideal types, exploring how the perfect construction should be, according to his idea of architecture (Cavalieri 2002). The ideal basilica he proposed was therefore a true building type, open to different interpretations. Vitruvius raised, in fact, the problem of the relationship between the reading of built reality and the architectural practice, still relevant and unresolved today (Gros 1984).

Saverio Muratori himself, to give an example closer to us, was not so much 'Muratorian' in many of his buildings, starting with the well-known housing intervention in the *Tuscolano* quarter of Rome, where the formation process of the Roman multi-family house, which he had studied for a long time, seems completely contradicted by the architectural product. Gianfranco Caniggia highlighted, in an exemplary way, the contradictions between design and theory in the work of Muratori (Caniggia 1984).

On the other hand, the best-known Italian theorists who have dealt with the problem, such as Aldo Rossi or Carlo Aymonino, although having conducted seminal studies of urban morphology on European cities, have always stated that the project cannot derive from the reading of the built landscape, having its own autonomous mechanisms; the reading of cities and fabrics has a general cognitive function, it is not a design tool.

I am convinced, however, that the project should derive from the reading of the built reality, from the analysis of the formation processes of fabrics and building types. I believe that this method contributes to the civil role of our profession and that it is our duty to 'insert' what we are building today into the flow of transformations of city and territory. Precisely for this reason, we cannot imitate the past; we are obliged to be contemporary. Our project, our constructions, are at the end of a process. Our personal aesthetic contribution, which can only be personal and original, is therefore a provisional synthesis of an ongoing development, even if we have to accept, it must be said, the contradictions and the uncertainties of a profession that contains, despite all the efforts, a certain unavoidable degree of empiricism.

I will use the Terni Cemetery, that I have been designing for years, as an opportunity to briefly reflect on some of my beliefs regarding the relationship between morphological reading and architectural design, at different scales, starting from the urban organism to the construction and language problems. The theme of the cemetery seems to me particularly congenial to the subject in question, constituting, in my interpretation, a city analogous to the real one, a sort of laboratory that contains the problems of morphological analysis, of architectural language and (if I am allowed to use the term) of poetic interpretation of the theme (Fig. 7.1).

Fig. 7.1 Early study sketches of the perimeter wall

Like any buildings for public use, the design of cemeteries must follow some general principles which, in architecture, as in every cognitive and technical activity of man, can be expressed through dyads of complementary notions, which form an important part of the substance inventiveness of architecture: the act of recognizing them in the reality of the many burial places where piety, the cult of memory or the admiration for exemplary lives have sown in every place of the civilized world, constitutes itself, not only a reading tool, but also a creative design operation. The forecast of future development therefore derives from the critical and innovative reading of the generative laws of the organisms of the past.

The project is thus based on two dyads of complementary notions derived from the critical reading of the formative process of the modern cemetery.

The notion of 'organism', as a relationship of necessity between elements and structures linked by a common functional, constructive and symbolic purpose;

the notion of 'congruence', related to the previous one, as a ratio of proportion and conformity of the elements that form the organism at different scales;

the notions of 'enclosure' and 'covering', associated with the basic gestures of inhabiting and protecting the space that coincide with symbolic forms, meaning by this those forms that reunite some fundamental aspects of knowledge in a synthetic and universal way, providing a conventional expression;

the notions of 'hierarchization' and 'specialization', linked to the previous ones, which translate into a code which, within certain limits, is universal. The formative processes of the enclosure, in particular, are linked to the concepts of centrality and periphery, of nodality and antinodality which constitute a fundamental key to interpreting the characters of architectural and urban organisms (Strappa 1995).

7.2 The Forming Process

The architecture of cemeteries is perhaps the part of our discipline in which, starting from the last century, the changes produced by industrial civilization, by the modern idea of progress and by the process of rationalization of the European city, appear most evident. Even in the second half of the nineteenth century, positivist thinking was opposed to the idea of expelling cemeteries from cities, claiming a fundamental civil role in burials.

In France, where the subject was tackled earlier and with greater attention than in other countries, the positivist philosopher Pierre Lafitte reaffirmed its role as an irreplaceable secular institution in civil life. Intending the scientific spirit as open to all the needs of the individual, Lafitte argued the necessity for symbols that recall what has been lost: 'the grave, he wrote, develops the sense of continuity in the private and the cemetery the sense of continuity in the city and in humanity' (quoted in Ariès 1977).

However, according to the ideas of Baron Haussmann and his officials, on the contrary, the large city cemeteries were destined to be transformed into simple functional infrastructures. Their project was to concentrate the burials in a single cemetery of 600 hectares, located 28 km from Paris, connected by a railway service to 'funeral stations' distributed throughout the capital (Etlin 1984).

Although Haussmann's project was never realized, it marked an important stage in the formation of the building and urban types of the twentieth century. The modern city reduces the problem of death, on the one hand, to private mourning and, on the other, to a simple public service: it is no coincidence that the cemetery will be the only major architectural theme removed from the research of Modern Movement, even if there are important exceptions in the first quarter of the twentieth century, as the Woodland cemetery in Stockholm by Asplund and Lewerentz (1915), the crematorium in Brno by Wiesner (1925–30), and the Northern Cemetery of Hilversum by Dudok (1929).

Even the fundamental notion of enclosure, of a functional and symbolic perimeter of the place intended for burials, changes according to the new roles attributed to cemeteries. There are two extreme examples. In the Cemetery of Pisa, 1270, the enclosure is interpreted as the boundary of the space reserved for the elect (Curl 1980). In this case, the image of the cloister suggests the protected sacredness of the place. The *Camposanto di Pisa* is in some ways also an urban museum, as evidence of continuity with the past, due to the ancient tombs reused by the Pisan nobles, which establishes a silent bond with the classical Christianized tradition.

Built six centuries later, the Vienna Cemetery testifies the transformation experienced by funeral spaces: it is also enclosed, but here the separation from the outside world takes place by a fence surrounding a new city similar to that of the living, borrowing the same contradictions and segregation mechanisms of the modern

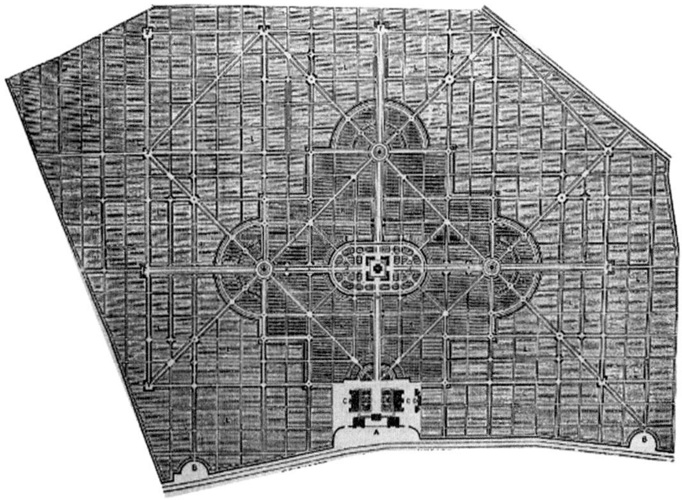

Fig. 7.2 Plan of the Central Cemetery of Vienna (*Zentral—Friedhof*). The affinity with a nineteenth-century city, of which it reflects the contradictions, is evident

metropolis (Strappa 1989a)—Fig. 7.2. The structure, organized by axes, nodes and poles, and designed as an urban fabric, is made up of three concentric areas distinguished by classes of burials. In the central pole the main chapel is located, surrounded by important funeral monuments (first class tombs); the second area is intended for private burials (second class tombs); the third to the mass graves (third class tombs). With the intention of giving more dignity and decorum to the cemetery, the main routes of the mass graves area are flanked by second class tombs, and those of the area with private burials, by first class tombs.

Despite the decline in devotion to the cult of the dead, the nineteenth-century cemetery remains, in many cases, a strong symbolic force also linked to secular and *bourgeois* values.

The functionalism of the modern city has deprived, in Italy, the cemetery of this civil role as well; the building image of the cemeteries of our cities, developed in the years after the Second World War, re-proposed the degradation of the urban suburbs of which, in this, it imitates the residential fabric.

It was only in the 1970s that the cemetery was renewed as an important architectural theme, with two exemplary buildings designed by Aldo Rossi (Modena Cemetery, 1971–78) and Alessandro Anselmi (Parabita Cemetery, 1967–82) as spaces intended, above all, for the living as small towns of memory (Anselmi 1986). Although these interventions do not derive from the interpretation of a process, their role in the renewal of Italian architecture was crucial.

7.3 The Fabric

In the past, those who had been part of a religious community or a civil context were buried *ad santos et apud ecclesiam*, inside churches, close to sacred places: they still belonged, in some way, to public life through remembrance.

The first fundamental fact that presides over the unitary formation of the specialized characters of modern structures (thus transforming what was defined as an 'open series' into an organism) is, therefore, scale: while the parish cemetery occupied a limited area, the modern one is a structure organized at an urban scale.

For this reason, the Terni Cemetery is based on the analogy of behaviour of its components with the urban organism: on the idea of the cemetery as a 'confined place' and, at the same time, a place that intimately belongs to the city and to its history (Figs. 7.3– 7.5).

The idea of continuing the Terni walls that separate the city from the countryside (the enclosed and protected part of the world inhabited by men from the vastness of the external territory) also intends to indicate the return to the belonging of the world of dead to the inner part of the enclosure, claiming the right of the memories to remain on this side of the border. It is also true, however, that the contemporary city, and perhaps the city in history, is not a safe place, least of all for peaceful rest.

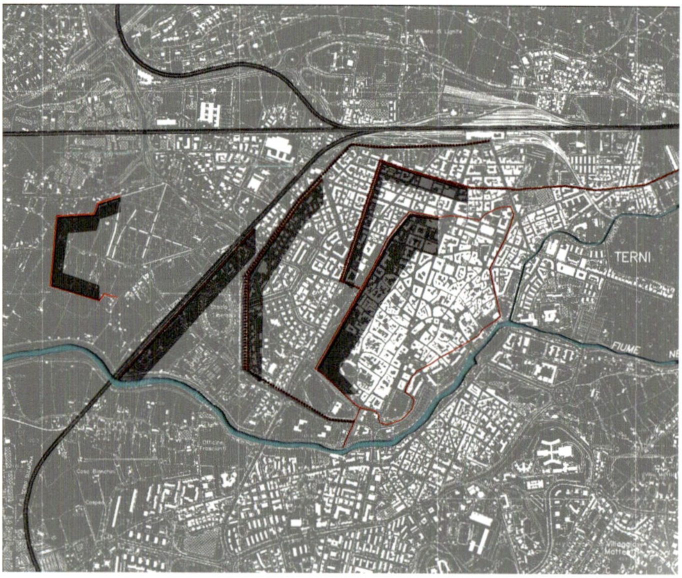

Fig. 7.3 The cemetery constitutes a new urban expansion limited by a further fence, a boundary between city and countryside. The new urban fringe continues the previous ones, from the Roman city walls to the railway line

Fig. 7.4 Model of the competition project

Indeed, it is a place of conflicts and contradictions, while the Umbrian countryside preserves the traces of a millenary agricultural civilization, which our memory, or perhaps our imagination, somehow associates with the serenity of peaceful work in the fields. This image is still very vivid along the *Marattana* route, close to the cemetery area, evoked by solid farmhouses, cultivated land and tree-lined paths. For this reason, the image of the new Terni Cemetery also contains traces of the rural past of the agricultural land transformed into a burial place. It is both a protected garden and a piece of the city (Figs. 7.6 and 7.7).

Fig. 7.5 Current situation (*Source* Google Earth)

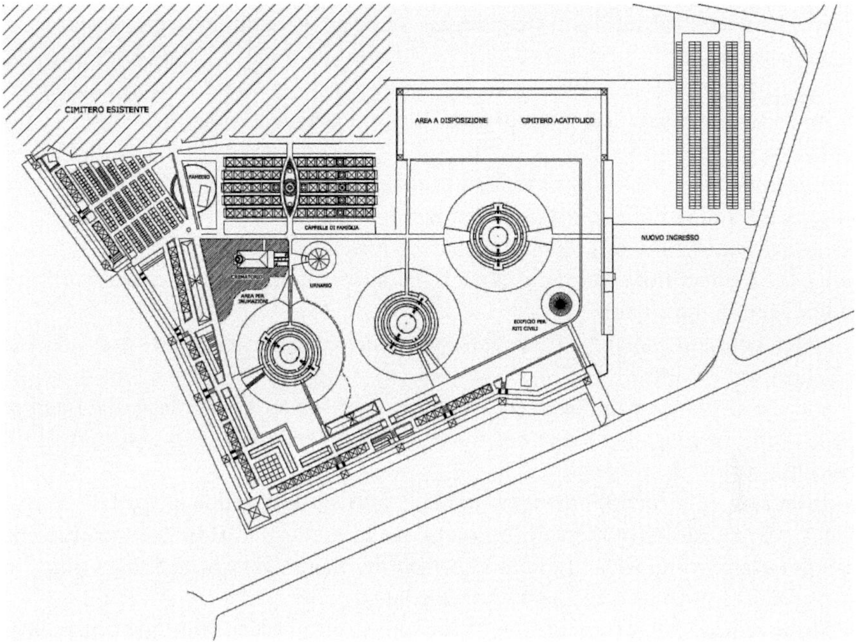

Fig. 7.6 Plan of the entire intervention

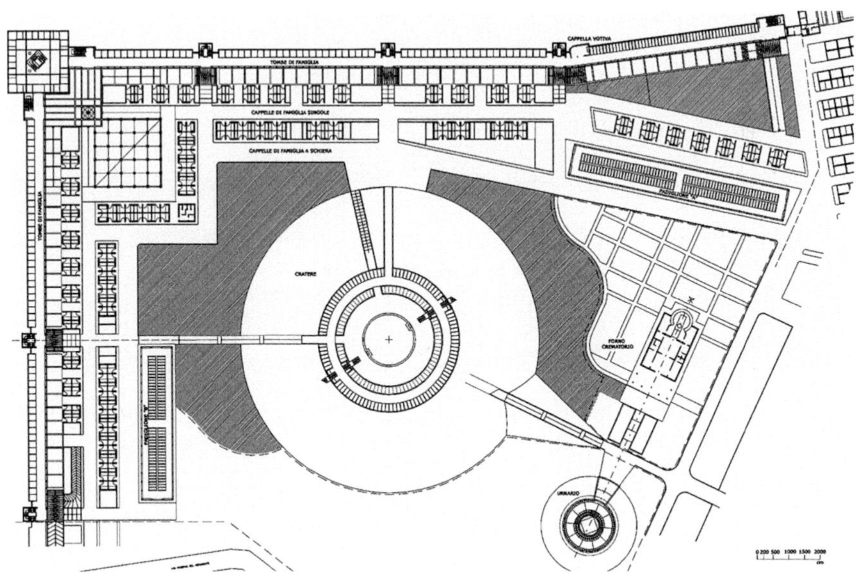

Fig. 7.7 Plan of the second (…) phase—works (…) carried out so far

For these reasons, the Terni Cemetery is made up, in addition to green spaces, as an urban fabric of:

routes, formed by arcades on two levels (to which, in some sections, an underground route is added) and external paths hierarchized according to the logic of the enclosure;

base buildings, consisting of the tombs that line up along the paths;

special serial buildings, consisting of family chapels;

special nodal buildings, consisting of the tower for the congregations (already built), the crematorium and the chapel at the new entrance to the cemetery (which will be built in the coming years);

nodes, consisting mainly of stair towers interpreted as vertical routes that intersect with horizontal ones;

poles, consisting of the intersections at a larger scale (the entrance, the internal space of the congregation tower and the crater at the end of the axis that starts from the entrance).

As in any architectural organism, in other words, the functional, spatial, expressive, static, constructive hierarchization of the elements and structures tends to be legible through volumes and routes generated by the movement that takes place in space, and the polarizing elements that regulate it.

The organism that is formed therefore assumes, on an urban scale, also the typical character of a large specialized organism (Figs. 7.8–7.10).

Fig. 7.8 Internal facade of the perimeter wall in correspondence with the part on three levels (*Source* photograph by A. Bravini)

Fig. 7.9 The nodal building at the intersection of the main routes (*Source* photograph by A. Bravini)

Fig. 7.10 External facades of the perimeter wall. In right, the stair tower awaiting the continuation of the wall (*Source* photograph by A. Bravini)

7.4 The Tectonic Sequence: Matter-Material-Element-Structure

One of the fundamental problems, from a morphological point of view, in the Terni cemetery design, was to identify the ways of transforming nature into architecture (Strappa 2014): how matter becomes material; how the material is transformed into an element of the construction; how the element contributes to the formation of structures, in increasing degrees; and, how the structures finally contribute to the formation of the whole architectural organism as synthesis and conclusion of a continuous process of transformation.

The choice of using plastic-masonry structures (Strappa 1995) was due to two categories of considerations. The first was linked to the notion of 'cultural area'. All the buildings in the Umbrian area have strong stone features that constitute one of the most vital factors in the cultural continuity of the civil communities that have transformed this territory over centuries. The wise use of stone is a choice still operating at the very heart of modernity: few regions have so jealously guarded their constructive heritage in our century and have updated and innovated it.

The example of Mario Ridolfi, a great architect from Terni by choice, was often cited as a modern continuer of a work linked to the sophisticated local culture of stone, a description that could be misleading in this sense. In fact, he ended up hiding, with the originality and substance of a long critical research, the permanence of a deeper, collective adhesion to the local roots. If the term had not already given rise to infinite misunderstandings, we could speak of a 'spontaneous consciousness' still operating.

The second consideration is the organic character inherent in the use of this material, the necessary continuity that is established inside the masonry walls and one that regulates the position of the elements in a stable way: independently of the innovate purposes, a good masonry building is linked to physical laws that impose an organic hierarchy between the parties (Strappa 1995), differentiating, in the case of the Terni Cemetery:

the base part (massive, opaque, load bearing and closing spaces);

the elevation (massive, opaque, load bearing and closing);

the unification band (consisting of the shaped concrete beam that supports the cast iron rods); and

the conclusion of the metal cover structure (light, transparent, load bearing and non-closing)—Fig. 7.11.

The use of stone has raised the eternal problem of cladding, as old as architecture. The walls of the building are plastic in nature (reinforced concrete walls), but the expression of this character is entrusted to the stone cladding.

Over the course of history, the use of stone has often involved the adoption of an 'indirect readability': the formation of tectonic nodes, of constructive solutions typified in the construction, have become a code to be used even when those nodes and solutions were no longer constructively evident.

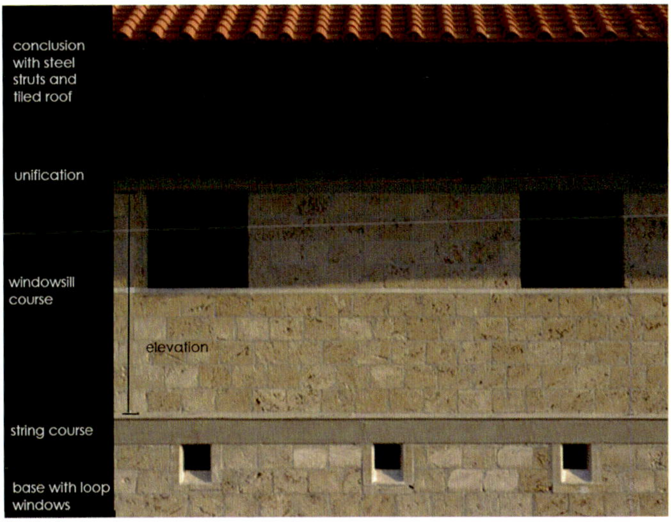

Fig. 7.11 The four zones of architectural stratification in the external facade of the perimeter wall (base, elevation, unification and conclusion) and some tectonic nodes. The windows of the base are of the loop type; the upper ones are formed by the void between two walls

Fig. 7.12 Interior of the stair towers. The elastic nature of the structure (load-bearing and non-closing spaces) and the filling function of the brick walls are highlighted

This is not a positivistic interpretation of architectural expression. The civilizations that produced the great codifications have always linked the material culture of the builder to the aspirations of the civil context: not the culture of the artist, individual, subjective, but that, more stable and authentic, of the craftsman.

Only in some historical phases, as in the gothic cathedral, do the external frontages represent the hierarchy of the interior spaces and the constructive needs of the building. Of course, this necessity, which presides over the formation of the elements that make the medieval organism so readable, has met with the favour of rationalist theorists, but in reality it constitutes, in my opinion, anything but the norm.

Today, the problem must be faced in its entire complexity and articulation. The modern idea of an architectural organism, from Leon Battista Alberti to Gottfried Semper, privileged the role of the facade as an 'indirect' synthesis of the building; the Albertian idea of elevation did not tend to represent the constructive *datum*, but to indirectly express spatial contents. In this sense, Palladio' s churches represent an extreme experimentation with indirect readability, where the facade becomes a description of a virtual space. The legacy that, as moderns, we collect from the history of architecture, therefore amply legitimizes the use of stone as a material that indicates its original load-bearing function without having a structural role.

In the case of the Terni Cemetery design, with the use of *sponga* (sponge) stone, the problem is simplified by the fact that this material must necessarily be used according to a collaborative role in the stability of the building, but not with a load-bearing function, due to its mechanical characteristics and its relative heterogeneity. A shuttering function was thus provided for the external stone walls, entrusting this material with a dual constructive and aesthetic function. Even if the construction of the building and the economic reasons did not allow the full application of this principle, the collaboration between stone, used in the dual role of cladding and building element, and concrete (a material that can also be used in a plastic method) is a fertile way forward (Figs. 7.13–7.15).

7.5 The Language Issue

What architecture literature usually means by 'language' is actually the set of tools through which a designer individually expresses his own works. Critics research and praise the originality and innovation of these tools. It seems to me an incongruous way of understanding the term 'language' in architecture. Any type of expression in architecture could be regarded as a sign system as in natural language, but in fact, the continuous, rapid, personal and radical transformations of the signs do not allow the formation of a code, the basis and foundation of any true communication.

Instead, I believe, we should recognize in the term 'architectural language' the value of personal use of a common and shared language, which should be understood, studied and respected (Purini 2011). There is no personal expression, in natural language, which should not refer to the deep structures of a common language.

Fig. 7.13 Upper gallery. The exposed concrete walls support the structure of the *polonceau* trusses bared by steel connecting rods (in the first phase these rods, intended to withstand only compression stresses, were made of cast iron; the new legislation practically prevents their structural use)

In this framework, as a literary work 'individualizes' (that is, it makes individual, unique and unrepeatable) some characteristics of the common language, a work of architecture should identify common features of a building culture, of which some shared characteristics are recognized. Qualities that, however, are updated, as with the written or spoken language, through the contribution of neologisms and syncretisms, or even through the 'high' experimentation of literature.

Expressed in these terms, the problem would seem solvable through the return to the founding matrices of language, to the common and recognizable forms of our architectural heritage. And yet, the problem is not so simple, as we cannot ignore the fact that a real language has been lost, that the meaning of the words has been either consolidated with use or has been forgotten. The only form of 'salvation', I believe, is given precisely by the 'materiality of architecture', by the order that the matter/material establishes and which is transformed into an order, or sign system with which the author communicates to the correspondent.

Architectural language, in fact, does not correspond entirely, although it has some similar characteristics, neither with natural language, nor with artistic language. Unlike spoken or written language, architecture not only indicates, marks, evokes reality; architecture 'is' reality. This explains why in periods of great crisis in architectural *langue*, we return, even in full modernity, to the concrete, material, constructive foundation of our profession. Think, for example, of the various research works of

Fig. 7.14 The upper gallery at the node of the stair tower. The cover is in exposed corrugated sheet metal (*Source* photograph by A. Bravini)

Fig. 7.15 The interior of the tower of the congregations, also containing the ossuary. The material used for the interior is the corrugated sheet used for casting the concrete of the floors. The use of steel in interiors aims to reflect the light that comes from below (*Source* photograph by A. Bravini)

Fig. 7.16 Internal passage (*Source* photograph by A. Bravini)

William Morris, Camillo Boito and Gottfried Semper. It seems to me that, especially in the current condition of globalization in communication, a contribution to the solution of the crisis in which the contemporary architect is struggling, could be provided by the recovery of the concrete and material aspect of architecture.

In this sense, it seems to me that even the reference to Mario Ridolfi, as previously mentioned, has the risk of being misunderstood as one of the many possible personal languages that crowded the architectural panorama at the end of the twentieth century.

In the Terni Cemetery design, I have tried to identify which characteristics were still operating in the specific building culture of the pertinent cultural area, bearing in mind that the productive reality, the customs and the 'typicality' of forms are linked to the constructive data and become understandable when deriving from that.

In the era of globalization, the very notion of 'cultural area', however, is being transformed. But, although it may seem the opposite, it is being extraordinarily strengthened. Just think of the continuity in the wooden elastic characters in the light and transparent structures and systems used in the high-tech forms, widespread, not surprisingly, in areas once linked to Gothic culture. In this context, it must be considered the modern reference to the specific ethos of the local language, which here is above all plastic-masonry.

Compared to the common stone character of the Umbrian-Tuscan area, the Terni area has some specificities that I have only fully realized when the detail design of the cemetry has started, after many visits to the modern areas of the city. In Terni, a city of industrial traditions linked to steel mills, the irruption of modern metal technologies has produced significant innovations. These have been consolidated for some time and, although largely not adopted for base buildings (the way of living has given rise, in every corner of the earth, to the most conservative building types) they are now part of the building customs of the place, especially for special buildings.

Furthermore, many public buildings in Terni, now partially demolished, were affected by the industrial character that had been added to that of the ancient city and adopted it, especially in the roofs, as a distinctive element. It seems that here the industrial landscape, the urban fabric and the rural world have found, since the origins of the nineteenth-century transformations, an unusual way to merge or, at least, to coexist.

As a consequence of these reflections, the fourth band of architectural stratification, the 'conclusion' that overlaps the plastic wall structures, is of a elastic character. The cast iron connecting rods are only used to support the overhangs of the *polonceau* truss roofs; the interior of the roof is made of corrugated sheet metal. Even the stair towers have a metal supporting structure, while the brick curtains have the purpose of closing the spaces. This choice corresponds to the desire to make the vertical path of the towers legible and avoid the risk of imitating the old fortified walls: the tower is not, here, the stronghold of a defensive structure, but the light interval, which uses a load bearing and non-closing, virtually transparent, serial construction system within the *sponga* stone curtain wall which, on the contrary, shows its massive, opaque, continuous, space-closing character at the same time.

7.6 The Aesthetic Synthesis and the Future of the Project

The uninterrupted flow of transformations is the natural condition, I believe, of the built world. It is necessary to take note of this: nothing is stationary; the shapes of buildings change like their physical substance. Buildings are formed by matter, they are transformed over time and return to the state of matter as a ruin. Architectures should be read and thought as an unstable part of a process.

In the construction of the Terni Cemetery, these transformations have continued until now, over a quarter of a century, with continuous, small but important rethinking, innovations and updates.

Despite the passing of time (debates, fashions and technologies), I have never believed that a radical makeover was necessary. To be honest in architecture, a non-progressive science, the answers to our problems have largely already been given. Recognizing them, today, in the age of praise of uncertainty and fragmentation, grasping their unity in the differences of current conditions, I believe is a true form of innovative and contemporary designing.

In the Terni Cemetery, this instance of unity, propitiated by a theme that asked for a high and quiet solution, seems to me to be captured in the passage from the stone walls of the first buildings, erected in 1996, on which the patina of time is now depositing, to the new buildings, where the new *carparo* stone introduces a new light and slightly different proportions. There is no evolution. Only a way of seeing the same things with different eyes, in the light of subsequent requests, of further reflections.

The Terni project, started from a didactical opportunity, from the students' unexpected passion for a difficult poetic theme such as death, which brought together a large group of designers for an 'academic' participation in a competition that I would never expected to win (Strappa 1989b, 2006). Since then, the project and its realization, have had ups and downs, between lightning-fast requests for variations and biblical waiting, with different construction firms—some competent and others not.

The phase illustrated in these pages, the most recent, was also the most fortunate, with municipal technicians unexpectedly attentive to architectural problems and a good construction firm that has finished the work earlier than required.

After so many years of practice with their shapes, I can no longer evaluate the emotions that these constructions can cause. A sculptor and friend who visited the construction site just after the finishing of works, used the expression 'severe grace'. I like the term 'severe'; it seems appropriated. It is a design, in fact, carried out without the romantic impulses that the theme could easily induce. It does not seem to me that there is the melancholy of detachment, nor the pain of lost affections. Rather, I hope, we perceive a quiet architectural *pietas* understood in the classical sense of the term.

The construction of a small votive chapel, however, took place under the wave of a strange emotion. I spent whole days redesigning the back wall, behind the altar. I looked for the glasses to be inserted, tested them in the light that was already penetrating by the structures under construction, personally set them on the concrete cast on a horizontal formwork washing it with a jet of water on the barely solid surface making the aggregates emerge (Figs. 7.17 and 7.18). When the wall was raised and mounted under the blade of light, I felt an indefinable sense of guilt for the irruption of a radiant and almost profane element in the religious half-light of the

Fig. 7.17 The votive chapel, inserted in the series of family tombs on the ground floor. As in many fabrics, the special building is obtained from the fusion of basic building units. The volume of the chapel protrudes from the external wall allowing the entry of light. The entire envelope is in concrete cast on site, except the wall behind the altar, which is casted horizontally

Fig. 7.18 Detail of the votive chapel

portico. I then convinced myself, almost by force, that this flash in the semi-darkness was logical and necessary.

A new construction cycle will begin in a few months. Changes requested by the municipal administration force the general idea to deal with the concreteness of practical and economic problems. Then I will have to define the final shape of the other, conclusive phase, including the orthogonal 'arm' which will conclude the large enclosure.

It is necessary to resist the temptation of change, the opportunity to 'keep up with the times'. There is no need. I talked about it with Sandro Anselmi, who wrote a generous article about my work (Anselmi 2012). He himself, an indefatigable experimenter, confirmed my idea of continuing the long shadow gap of the portico, perhaps by other means, but with the same purpose, with the same unifying rhythm.

This harsh self-censorship, in the end, will be rewarded by the erection of the same inhabited wall over a kilometre long, a huge sign on the territory where only a few differences, which will escape the sight of the visitor, will mark the slow passage from one construction season to another. It will take even more time. Maybe it will all be over in another quarter of a century.

Technical Data

The final project for the extension of the Terni Cemetery was drawn up by the design team G. Strappa (group leader), T. Casatelli, P. Di Giuliomaria and E. Timpani. The project is linked to a long technical-bureaucratic process that took place over many years, starting with the National Competition launched in March 1986 and won by the project indicated by the motto 'The Good Earth' designed, in its early version, by the team G. Strappa (group leader), T. Casatelli, M. Capotosti, P. Di Giuliomaria, M. Pagotto, M. Pisani and E. Timpani. The first working plan was delivered to the Administration in December 1990 and the final variant in July 1996. The structural calculation was carried out by prof. Antonio Michetti. The first phase of the entire programme was carried out in various periods: the first by the firms Baldassini Tognozzi and Cadam firms, for a volume of 16,000 c. m. and a cost of € 5,800,000, was completed in 2011; a second by the firms Gefim, Celi, Consorzio Cooperative Costruzioni for a volume of 38,000 c. m. and a cost of € 14,000,000, was completed in 2018. At the beginning of 2021, the construction of a further phase on an area of approximately 20,000 m^2 will be tendered. All the project phases were carried out with the collaboration of the Municipality of Terni, Public Work Department, whose current general manager is arch. P. Giorgini and technical manager Ing. L. Donati with the collaboration of Surveyor G. Poddi.

References

Anselmi A (1986) Vent'anni dopo, alcune riflessioni sul cimitero di Parabita. Verona, Architettura, Monumento, Memoria

Anselmi A (2012) An architecture of the shadow. The new Cemetery of Terni by Giuseppe Strappa. Paesaggio Urbano/ Urban Design 3:20–31

Ariès P (1977) L'homme devant la mort. Edition du Seuil, Paris

Caniggia G (1984) Saverio Muratori e il progetto di tessuto. Storia dell'architettura

Cavalieri M (2002) La basilica civile nel de Architectura di Vitruvio: prassi e codificazione in Italia e a Parma. Arcihivio storico per le province parmensi

Curl JS (1980) A celebration of death. Constable, London

Etlin R (1984) The architecture of death—the transformation of the cemetery in eighteenth-century Paris. MIT Press, Cambridge

Gros P (1984) La basilique du forum selon Vitruve: la norme et l'expérimentation. Bauplanung und Bautheorie der Antike, Berlin

Purini F (2011) Un'architettura tra origine e inizio – riflessioni sull'architettura di Giuseppe Strappa in occasione della seconda fase dell'ampliamento del cimitero di Terni. Ar 95

Ragon M (1981) L'espace del la mort. Albin Michael, Paris

Strappa G (1989a) I cimiteri. In: Carbonara P (ed) Architettura Pratica II. Utet, Turin

Strappa G (1989b) L'ampliamento del cimitero di Terni. Industria delle Costruzioni 5

Strappa G (1995) Unità dell'organismo architettonico—note sulla formazione e trasformazione dei caratteri degli edifici. Dedalo, Bari

Strappa G (2006) Ampliamento del cimitero di Terni. Oltre l'architettura moderna, Quaderni di Ajòn, Florence

Strappa G (2014) L'architettura come processo—il mondo plastico murario in divenire. Franco Angeli, Milan

Chapter 8
Rigour and Respect for the Urban Context

Matteo Ieva

Abstract How to consider the project intervention in an urban context which is conditioned by the existence of special buildings that express a significant historical character? Against a widespread tendency that considers his own critic operate of living the contemporaneity as an expression of the 'being' as a revelation of his own time, a school of thought is outlined, proposing the designer's critical action through a proactive research field that subordinates authorship in favour of an intentionality conditioned by the constitutive phase of the 'being' within which to intervene. It is to this second research line that the project of the in-line building for commerce, offices and apartments in *Canosa di Puglia* refers to. It reads reality not as a temporal instant, but as a process that has in itself the law of its unfolding. It pursues a perspective of transformation with a critical horizon based on the concept of historical continuity and of an organism made up of parts, congruently connected to each other. The building is on a plot located on the north corner of a small street-block. It attempts to establish a complementary relationship with the surrounding area, especially with the historical 'emergencies'. The renunciation to an aesthetic readability of exaggerated and useless intentionality is based on a programmatic desire to consider the new intervention as a 'temporary lacuna' of the existing urban fabric. Therefore, the recovery of the autochthonous character and the adherence to the models of behaviour of the local constructive tradition have suggested a set of design choices aiming at privileging the 'environmental value' and denying an attitude of solipsistic alienation from the context.

Keywords Canosa di puglia · Cultural area · In-line house · Architectural language · Aesthetics · Identity

---

M. Ieva (✉)
Polytechnic University of Bari, Bari, Italy
e-mail: matteo.ieva@poliba.it

V. Oliveira (ed.), *Morphological Research in Planning, Urban Design and Architecture*, The Urban Book Series, https://doi.org/10.1007/978-3-030-66460-2_8

8.1 Theoretical Principles and Methodological Questions

> The closer we get to danger, the more the ways towards what saves begin to light up, and the more we ask. Because asking is the piety of thought. (Heidegger 2017)

To introduce the topic of this chapter, I would start with a first question: how to consider the design intervention in an urban context conditioned by the existence of special buildings, expressing a significant historical character?

From this interrogation arises another question related to the meaning of 'historicity' and, therefore, to the value attributed to past evidences in the built environment. It is our conviction that it is only by starting from the interpretation of the extant landscape that research trajectories of a project can be delineated and proposed, in a way which is coherent to 'that' reality, with its specific qualities (but also with the undoubted contradictions) expressing its 'being' as a testimony of civilization, in the moment of intervention.

An architectural project is a critical-individual act that can vary between two opposite poles in terms of relation with the built environment. The first is the search for a 'way of doing' that can aim at establishing a relationship of 'continuity'. The second aims at interrupting the process—and, therefore, it leads to 'discontinuity'. Interpretative expressions of the architect's work are included in the word 'project', as future projection. There are two possible meanings. One is founded on the Aristotelian *proairesis*, associated with the concept of project that contains the past in itself as a value, and that also prodigiously presents the Τύχη. The other is founded in the German word *Ent-wurf*, where the *Ent* preposition means detachment, often understood as detachment from the past and from history (Cacciari 1981).

This premise leads us to reflect on the different nuances in the field of activities to which the designer is called to participate and on the critical choice of establishing a relationship with history, contributing for assessing the importance of past evidences as a live expression of former civilizations.

To clarify the 'sense of historicity' from which to move in order to interpret 'the past that lives in the present' and defines the background for the intervention, by determining a dialectical parallel, we will try to define that meaning by recalling some theoretical positions that expand the critical view on this subject.

Gadamer (2000), for example, observes that the 'restored' past (see F. Schleiermacher 1996s hermeneutics) does not coincide with the authentic, and proposes the path of integration into the present's life. For this reason, he judges it as something that is 'alive' and continues—operating—to speak. From this point of view, operating in a context that offers traces of a significant antecedent indulges us to be part of the historical process, which must be interpreted as a synthesis of two moments: the 'transmitter' of the historical data, on the one hand, and the interpreter, with his mental mechanics grounded on historicity, on the other hand. Gadamer suggests the word *wirkungsgeschichte* (history of effects), calling attention to its inclusion in the historical process through the 'fusion of horizons'. That condition would allow evaluating its immanence as a living attribute that also participates in the present.

The Italian philosopher G. Agamben suggests that there is always something more intense to discover when one goes back to the moment of manifestation of a fact that has been handed over history (Agamben 2005). That is because it is only in a deferred time that something can appear in a dimension, where one can grasp a 'place for the possible in the past'. This place is called 'philosophical archeology'. It is the a priori of the past (a concept close to Foucault postulate) expressing an 'invitation' to examine the interrogation of the present casts on the past, trying to search for the *archè* (the first principles that make possible a condition of 'truth', that is not related with a chronological perspective). Agamben argues that this can be grasped in the gap between the onset point of a phenomenon and the tradition operating in the time in which it is judged. Therefore, knowledge reveals its way of 'being' and of manifesting itself as a product of a culture, variable in space and time in relation to the (historical) 'values' transmitted. It is in this line of judgement that the ethical need to establish a relationship of congruence to, and respect for, the inherited landscape is recognized.

The Muratorian-Caniggian school of thought argues for considering the critical action of the designer through a context of proactive research, subordinating 'authorship' in favour of 'intentionality' conditioned by the constitutive phase in which action is taken (Caniggia 2006; Caniggia and Maffei 1984, 2017; Muratori 1963).

Therefore, the context in which we operate, considered in the perspective suggested by Agamben, actively participates in the definition of the project and the choices to be made (typological, material, morphological and linguistic). If a high historical value is recognized, the designer's critical *cogito* does not 'announce' itself with an assumption of total adherence to the values in place at the time, but it seeks the link with history, by subordinating his choices to the character expressed by the urban context, in its more authentic meaning. This interacting relationship between subject and object, framed by the theory of the Muratorian-Caniggian school, offers a dialectical system that is established between the 'maker', with his beliefs and belonging to the present, and the 'built', with its limits but also its expressive virtues of the cultural and civil aspects (of which it is a direct manifestation).

This expresses the paradigm of an objective-subjectivity that looks at the built landscape and suggests its transformation without prioritizing the *ego*. The project that I am going to present makes explicit reference to the theoretical position just discussed.

8.2 The Building, Erected Between the Theoretical Framework and the Critical Design *Cogito*

The in-line building for offices and apartments in *Canosa di Puglia* is located in the street-block defined by the streets of *Sant'Angelo*, *Benedetto Brin* and *Santa Lucia* (Figs. 8.1 and 8.2). It has four storeys and basement for parking.

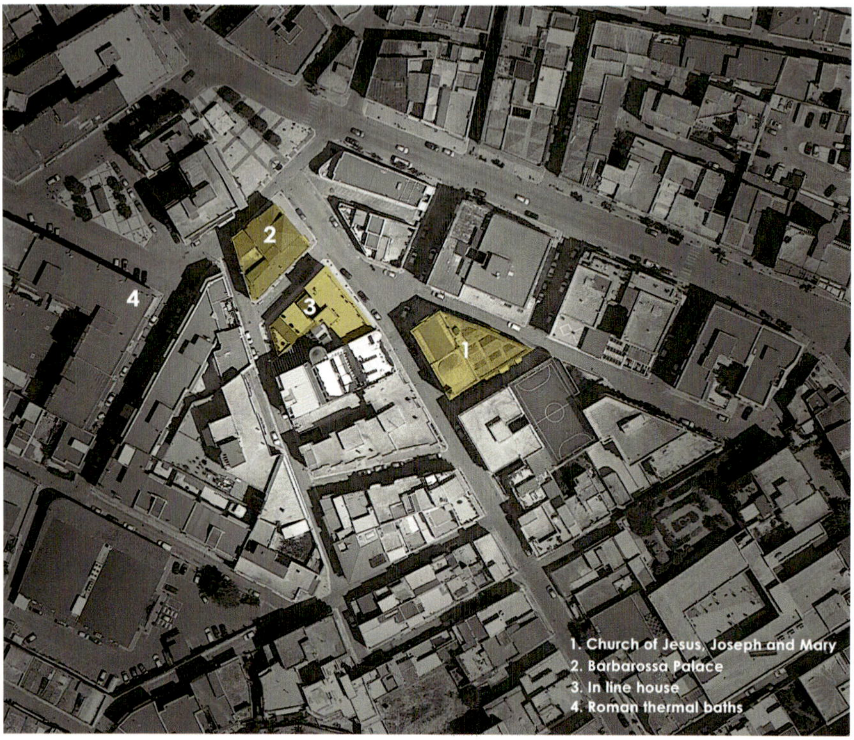

Fig. 8.1 The urban context around Via *Santa Lucia* (*Source* Google Earth)

The first storey, with a larger surface area, identical to the ground floor (currently the home of a banking institution), is entirely used for offices, while the second and the attic floors are both used for apartments.

The urban context in which the building is inserted has been erected between the late 1800s and the early decades of the 1900s. Testimonies of particular architectural value are the Church of Jesus, Joseph and Mary (nineteenth century), the *Palazzo Barbarossa* (the only example of a special building in Art Nouveau style, in *Canosa di Puglia*) and, very near, a *substratum* of Roman baths in the basement of a contemporary building (Strappa et al. 2003).

The surrounding fabric, except for some recent buildings that contrast with the urban context, is quite homogeneous; most buildings share a similar character.

The new building is in the north head of a small street-block and stands along an ancient path that connected an ancient Italic temple (later transformed into a Christian basilica) with the area of the Roman forum. It tries to establish a complementary relationship with its surroundings, especially with the aforementioned historical 'emergencies'. The renunciation to the seduction of giving the aesthetic readability of the building an exasperated, but useless, intentionality derives from the programmatic will to consider the new intervention as a 'temporary gap' in the existing fabric.

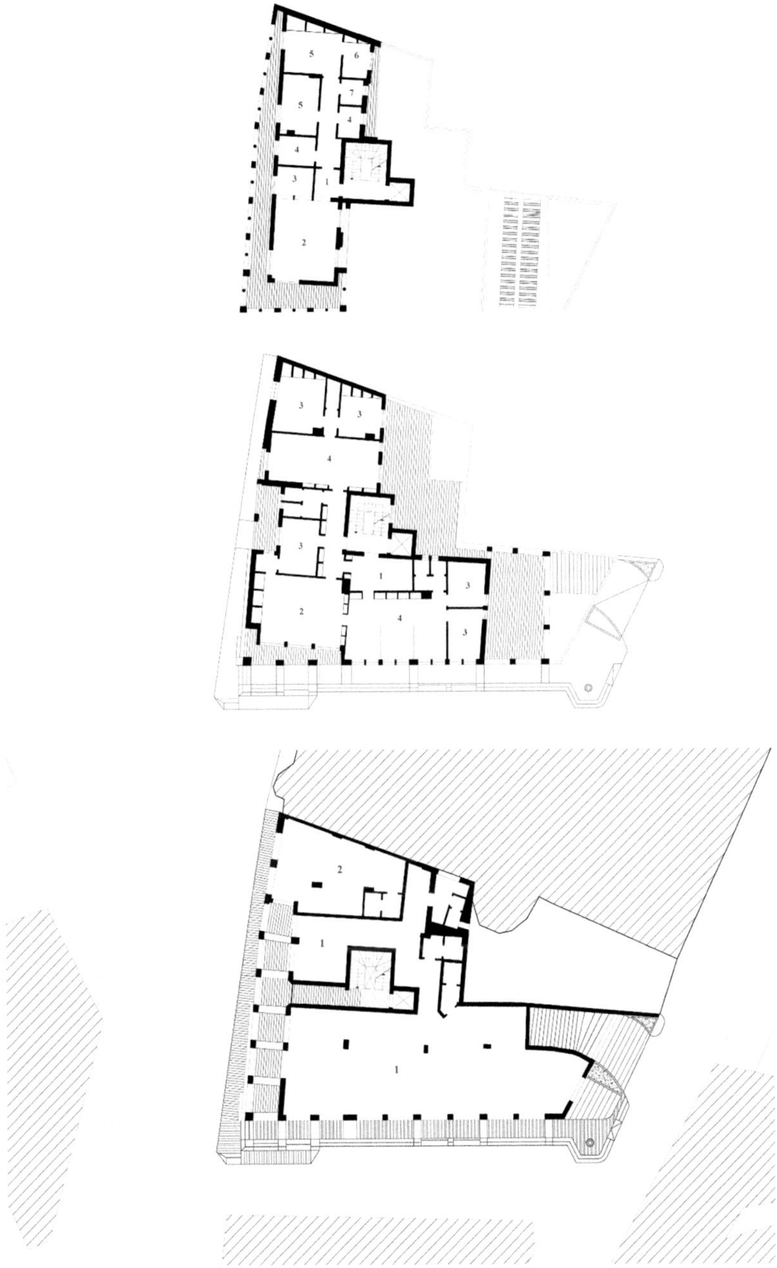

Fig. 8.2 Attic floor (**1** hall, **2** living-dining room, **3** kitchen, **4** bathroom, **5** bedroom, **6** Wardrobe, **7** closet), first floor (**1** hall, **2** meeting room, **3** office) and ground floor (**1** bank, **2** shop)

Therefore, the recovery of the autochthonous character and the adherence to behavioural models of the local construction tradition has suggested design choices that privilege the 'environmental value', with the consequent renunciation of an attitude of solipsistic estrangement from the context.

Despite the severe regulatory constraints, the building retains dimensional and proportional relationships to the historical permanencies, attesting its hierarchical subordination. That is the case of the relationships between the bands and cornices of *Palazzo Barbarossa* and the nodes of the in-line building, intentionally placed at the same height to establish a congruent relationship in volumetric terms (Fig. 8.3).

Regarding the design choices at the scale of the architectural organism, the appropriate use of the static system with reinforced concrete frame (load-bearing, non-locking) should be highlighted. Such a constructive choice—in addition to the convenience of the construction site for the erection of the structure—has allowed solving the regulatory constraints and adopting appropriate solutions, also in relation to the difficulty of founding the building in a place where there is a substratum of man-made cavities. For this reason, the solution which leads to an increase of the edge pillars (clearly legible in the façade, from the surface in contact with the ground up to the roof—Figs. 8.4 and 8.5) has been adopted. In fact, an ordered sequence of serial

Fig. 8.3 Via *Santa Lucia*: *Palazzo Barbarossa* and the new building (*Source* Photograph by Giuseppe Volpe)

Fig. 8.4 The corner of Via *Santa Lucia* and Via *Sant'Angelo* (*Source* Photograph by Giuseppe Volpe)

Fig. 8.5 Detail of the façade in *Via Sant'Angelo*

vertical elements appears clearly along the two external elevations, interrupted only at the level of the 'string course' between the ground and first floors and between the second and attic floors. A hierarchical stratification that intentionally recalls the tectonic sequence of buildings in the 'organic masonry' (*organico muraria*) geographical-cultural area, in which the different levels are distinguished in relation to the role they play within the organism: the base, a first level of contact with the ground; the elevation, a succession of serial planes; and the unification/conclusion, given by the attic floor and the frame system (Strappa 1995). This succession of differentiated levels recalls (by updating, and not imitating, it) the typical character of traditional Apulian architecture. Building with coherence and respect in the cultural tectonic-masonry field means proposing solutions that capture the 'essence' of the local architecture's character, openly organic and unitary. The purpose was not to step back from current construction materials and techniques but, to interpret construction framed by the principles and specificities of the cultural area.

The readability of the 'whole' has been thought in coherence with the 'type' of building (mixed uses), with the construction technique adopted, with the functional-distributive articulation, with the relationship with the context, with the aesthetic synthesis that harmoniously combines all components defining the architectural organism, but above all express a sense of belonging to a geographical area in which the 'identity values' of Mediterranean culture are recognizable (Strappa 2014).

But, how to recognize the identity character, so that it can be critically reinterpreted through the project?

Identity should be seen as the recognition of constant and repeatable qualities, taken as a value, which testify the essence of the building, as part of an architectural culture. This reflection also recalls the sense of the *eidos*, that is, the conscious search for the internal nature of things, as an analysis of those elements that make something to be what it is, enabling the understanding of its meaning. The Muratorian-Caniggian school, in this research on reality, proposes an analytical method based on the foundations of knowledge that allows to define this character, identified in its being co-essential, frequent and transmissible, useful for describing the differences that characterize the cultural sphere of which it is a direct expression.

Organisms (building, aggregative, urban) representative of a geographical-cultural context, interpreted as competing entities are studied in the internal relationships designating the complex of 'universal correlations' that characterize them, both as objects in themselves, and in correspondence with the other objects to which they are connected. Their structural-procedural variability suggests that even architecture (as any linguistic entity) has a community-based nature, of which the distinctive features of a shared architectural work ('spoken' and 'written') must be sought.

Once defined the general aspects of the problem, we can consider that to understand the aesthetic-identity essence of a place, it is necessary to critically reflect on its constitution and historical process, starting from the assumption that it has to be considered as a visible result, a sensitive semblance of an architectural and urban idea (according to Hegel 1997, 2000, semblance itself is essential to the essence). This consideration obviously stimulates the interest in correlating the aesthetic dimension (seen from the technical-practical point of view that takes up the ideal of the *pulchrum* as a structural-figurative expressiveness) with the ethical entity, deeply manifested in the 'way of being' in reality and in the expression of collective behaviours.

From these considerations (of a general nature), it can be observed that the aesthetic judgement (not contemplative or referred exclusively to a purely logical concept and to 'taste'), finds its own dimension in the tangible concreteness of the constructed reality, placed in the transmissibility of the *sensus communis* and kept in the transcendental condition of the 'intersubjective' relationship (with a premise of the search for truth, employing shared elements), as an indispensable presupposition of 'universal subjectivity'.

Among the values recognized as identifying the geographical-territorial sphere in which the city of Canosa di Puglia takes part, there is the role of the envelope defined by masonry systems, simultaneously bearing and closing, to claim the 'sense' of traditional massiveness (historically determined by the availability, and use, of stone) translated into the opacity of the surface, achieving also a natural sustainability, without the use of technical tools or technological expedients.

The legible *facies* of the building intentionally shows this character, proposing doors and windows in the continuous wall (allowing ventilation and lighting of the interior rooms). It should be noted that the northeast façade, along Via *S. Lucia*, has only a few windows and adopts a backward *loggia*, because at this latitude this orientation is avoided during the hottest hours of the day. The *loggia* in the façade of Via *Sant'Angelo* is due to privacy concerns and, at the same time, it deals with the lack of orthogonality of the different streets. To preserve the unity of the frontage (non-transparent building envelope) the office floor maintains the same layout as the openings of the upper living floor on Via *Santa Lucia*. In the other frontage, it adopts a more permeable surface (punctuated by the system of pillars) representing the role of the different parts, made of the structural elements and the wall (stepping back from the structure).

The attic floor, similar to the proposals developed by Italian modernity, includes a continuous *loggia* delimited by the perimeter pillars and by concrete elements, of circular section, at the centre of each space (Fig. 8.6).

The upper limit of the building, made of an extended slab supported by circular pillars and a small reinforced concrete partition, resting on the perimeter beam, reinterprets the traditional role of the final frame of special buildings offering protection to the façade (Maffei and Maffei 2018).

Regarding the 'work plan' (*piani di lavoro*) describing the general behaviour of all component systems—i.e. the *Sachlichkeit* that expresses the 'truth' which the organism aspires to—the external surface shows the role of the stuctural parts, constituted by pillars that are slightly extended from the building limit, articulating the rhythm of the different walls.

Research on the hierarchical role of individual elements also seeks a possible, and above all, integrated logic between the meaning of the external surface of the wall and the reference to the 'nodes'. In fact, Trani stone slabs, covering and protecting the masonry, are arranged in such a way as to mark (due to the different size of the elements) the level of the traditional *marcapiano* and the *marcadavanzale*. The result is a vertical partition structured in a way that is consistent with the double relationship

Fig. 8.6 The upper limit of the building (*Source* Photograph by Giuseppe Volpe)

that is defined in the dialectic tradition/modernity. In addition to the 'overlapping' of distinct floors, differentiated by role and function, there is an interest to declare a kind of constructive sincerity, by its nature close to the behaviour of the discrete, load-bearing and non-closing systems, characteristic of areas with an elastic-wooden tradition. It is precisely the use of the reinforced concrete frame system which, by its constitution, has a behaviour with repeated and serial elements, very different from the Mediterranean nature, that has developed continuous and non-transparent surfaces due to construction needs and thermal convenience, considering the high temperatures during the summer season.

8.3 Conclusions

In conclusion, generalizing the discussion on the problem of architectural identity and on the meaning of a cultural area, it is noted that a structural difficulty on the understanding of the horizons of reflection consistently pursued today to the topic of the essay, the perspective to which the design research presented here tends to be placed in a furrow of critical reflection that tries to outline an observation trajectory far from any speculative abstraction and deployed on a theoretical position that outlines an idea of architecture based on the idea of identity space (Purini 2008). Identity does not renounce the establishment of contact with other worlds and cultures. It declares a condition of necessary, but measured, contamination as a prerogative of existence in a contemporary age, which makes the area boundaries more fluid, without sacrificing the cultural essence.

The *Canosa di Puglia* building aims at identifying a hypothesis of profound rigour and respect for the urban context, pursuing a condition that is contrary to the identity nihilism of many contemporary projects. It offers a different architectural conception based on a form of 'critical internationalism' (Gregotti 2006), capable of preserving local lexicological distinctions, but also of recognizing the evidence of an architecture inspired by a 'new language' of European origin, in which experiments are framed by a wide limit that tends to make the differences hybrid. A direction into the future (*infuturarsi*) is favoured, through the profound bond that is established between our *Machenschaft*, our 'being', active and purposeful in transforming the future, and those ethical and aesthetic components that, in the field of architecture define, our concrete existence (Adorno 1975). In other words, we seek the ontological scope, our way of being in the relationship with others and with the world, in the search for an urban aesthetic that can express itself in a completely new founding and inaugural scope (Vattimo 2012): 'opening a world' which recognizes the unity (of the architectural language), despite the multiplicity of outcomes and a dialectically intersubjective vision—an inalienable prerogative of any linguistic entity.

Acknowledgements The author would like to thank Carmine Robbe, the co-author of the architectural project.

References

Adorno TLW (1975) Teoria Estetica. Einaudi, Torino
Agamben G (2005) Homo sacer—il potere sovrano e la nuda vita. Einaudi, Torino
Cacciari M (1981) Progetto. Lab Polit 1:88–119

Caniggia G (2006) Ragionamenti di tipologia. Operatività della tipologia processuale in architettura, a cura di G.L. Maffei. Alinea Editrice, Florence
Caniggia G, Maffei GL (1984) Il progetto nell'edilizia di base. Marsilio Editori, Venice
Caniggia G, Maffei GL (2017) Interpreting basic buildings. Altralinea Edizioni, Florence
Gadamer HG (2000) Verità e método. Bompiani, Milan
Gregotti V (2006) L'architettura nell'epoca dell'incessante. Laterza, Bari
Hegel FGW (1997) Estetica. Einaudi, Torino
Hegel FGW (2000) Fenomenologia dello spirito. V Bompiani, Milan
Heidegger M (2017) La questione della tecnica. goWare, Florence
Maffei GL, Maffei M (2018) Interpreting specialised buildings. Altralinea Edizioni, Florence
Muratori S (1963) Architettura e civiltà in crisi. CSSU
Purini F (2008) La misura italiana dell'architettura. Laterza, Bari
Schleiermacher FDE (1996) Ermeneutica. Rusconi, Milan
Strappa G (1995) Unità dell'organismo architettonico—note sulla lettura dei caratteri degli edifici. Bari
Strappa G (2014) L'architettura come processo—il mondo plastico murario in divenire. Franco Angeli, Milan
Strappa G, Ieva M, Dimatteo MA (2003) La città come organismo—lettura di Trani alle diverse scale. Adda, Bari
Vattimo G (2012) L'arte e la sua verità. https://www.youtube.com/watch?v=3rWZu3rECjg. Accessed 31 July 2020

Chapter 9
Contemporary Architectural Projects as Interpretation of Urban and Social Contexts

Gianpiero Moretti

Abstract The North American city has great potential for exploring the relationship between existing, relatively new, urban forms and contemporary projects. Often, contemporary projects are part of urban processes that are underway, even in central parts of cities. In this case, the analysis of the urban and typological evolution is possible thanks to the use of cartography, making possible to highlight the strata of formation and transformation of the urban fabric, which remains an open process for contemporary projects. On the other hand, being relatively recent urban forms, the link between these and the cultural and social context that produces them is more evident. This new perspective opens the door to a 'sensitive morphology' in which the contemporary project establishes a dialogue between the history of the city, and its architecture, with the aspirations of its inhabitants. It is in this frontier that the project discussed in this chapter was conceived. Located in Quebec City, Canada, this small property complex of three units completes a street corner in the *Montcalm* district. In the first part, we will briefly discuss the emergence of the neighbourhood and in particular the role of the plot. In the second part, the project will be addressed both as an element of the urban ensemble and as a cultural product related to the intentions of the owners at a particular moment in the history of the city. Finally, a brief discussion will underline the specificities of this contemporary intervention in an ancient North American fabric.

Keywords Urban morphology · Urban form · North American urban form · Contemporary architecture · Urban architecture · Quebec City architecture

9.1 Introduction

Since the emergence of the field of urban morphology in an architectural tradition, the relationship between the observations carried out in the city and the development of an architectural or urban project was at the centre of debate. The studies of Saverio

G. Moretti (✉)
Université Laval, Quebec City, QC, Canada
e-mail: gianpiero.moretti@arc.ulaval.ca

V. Oliveira (ed.), *Morphological Research in Planning, Urban Design and Architecture*, The Urban Book Series, https://doi.org/10.1007/978-3-030-66460-2_9

Muratori on Venice (1959) and Rome (1963) testify, as a manifesto, the concept of operationality of urban and architectural history (*operante storia urbana*) by putting forward the idea of continuity of the constitutive process. This results of observations on urban fabrics and building typologies throughout the history of the built context.

The links between history and project are taken to the extreme in the case of the transformations of the city of Bologna in the 1970s. In Bologna, the investigation carried out, in a rigorous manner, on the urban fabrics and existing building typologies allowed for a careful intervention on the whole historical centre (Cervellati and Scannavini 1975). If this approach seems appropriate for the transformation of existing buildings, important questions have been raised about the use of extant architectural types, derived from morphological studies, for contemporary intervention in the centre of the city.

On the other hand, more openly, in his proposals for *La città analoga*, Aldo Rossi (1976) theorizes the utilization of the extant city and its architectures for the design of the contemporary project. In this approach, the city becomes a dictionary and the project a mixture of desire, dream and reason. It was in the mid-1970s that Rossi's work was introduced to the United States, first with exhibitions and teaching, and later, in the early 1980s, with the translation of *L' architettura della città* which would have wide circulation in the Anglo-Saxon world (Rossi 1966). The work of Aldo Rossi, of easy reading (when compared to other Italian architects), would have a great echo and diffusion in the United States, where the idea of utilization of extant architectures would mix with a certain monotony of built form production. In this case, the historical references often boil down to stylistic quotes, which are superimposed on modern architectural forms in their essence.

However, the North American city has great potential for exploring the relationships between existing urban forms, which are relatively recent, and contemporary projects. Often, contemporary projects are part of urban processes that are underway, even in the central parts of cities. In this case, the analysis of the urban and typological evolution is possible thanks to the use of cartography, making possible to highlight the stages of formation and transformation of the urban fabric which remains an open process for contemporary projects. On the other hand, being relatively recent architectures and urban forms, the link between these and their cultural and social contexts is more evident.

In the case of Quebec, Dufaux (2007) underlined the link between the implementation of urban and architectural forms in various Montreal neighbourhoods and the cultural, social and economic context that produced it. This new avenue opens the door to a 'sensitive morphology' in which the contemporary project establishes a dialogue between the history of the city, its architectures and the aspirations of its inhabitants.

It is in this border space that the project under discussion in this chapter—designed in collaboration with A. Vallières—fits in. Located in Quebec City, Canada, this small building complex of three units completes a street corner in the *Montcalm* district. Firstly, we briefly discuss the emergence of the neighbourhood and in particular the role of the plot of the new building in the evolving urban structure. Secondly, the project is considered both as part of the urban whole and as a cultural product linked to the intentions of their residents and, therefore, to a particular moment in the history of the city. Finally, a brief discussion addresses the specificity of this contemporary intervention in a North American urban fabric.

9.2 Quebec City and the *Montcalm* Neighbourhood

Founded in 1608, Quebec City is characterized by a promontory site that opens to the St. Lawrence River. From its foundation, the geomorphology of the place of establishment has framed the division between the Lower Town, in contact with water as a place of production and trade, and the Upper Town, a place of defence and power. In the upper town, three parallel ancient routes structure the territory. Among these, the *Chemin Sainte-Foy* has guided the agricultural development, the development of the first suburban villas and the town growth to south-west in the northern part of the plateau (Fig. 9.1).

The project is in the *Montcalm* district located south-west of the historical centre of the city. The first construction of the district takes place in the *Chemin Sainte-Foy*, hitherto characterized by the construction of suburban villas. In the new district, the *Faubourg Guénette*, after the 1830s, a new building type started to be erected, the *Maisons de Faubourg*, typologically similar to those of the *Saint-Jean Baptiste* district. These small one-and-a-half-storey houses accommodated both French speakers and Irish families. Organized around *Candiac* Street, this small complex ends in the countryside bordered by *Cremazie* Street. The slow and almost spontaneous construction of the suburb continued for several decades. In 1875 about 40 plots were occupied by houses.

In an 1879 map of the parish of *Notre Dame de Quebec* (Fig. 9.2), we can see the grain of construction along *Chemin Sainte-Foy* as well as that of *Faubourg Guénette*, the only place with a continuous construction made of small Faubourg dwellings which contrast with the large villas of *Chemin Sainte-Foy*. In the same map, north of *Rue Saint-Jacques* (now *Rue Crémazie*), there is a subdivision route still unoccupied. These first attempts of planned construction of the sector continue with the plan led by the promoter Boivin. In 1874 he designs a subdivision of 300 units around the current *Rue Cartier*, between *Avenue de Salaberry* and *Avenue de Bourlamaque*. These attempts remain unsuccessful, due to the poor economic conditions and the lack of transport to relate this part of the territory with the historical centre.

The arrival of the tramway in 1898 (*Fronten/Des Érables* line) gave a new impetus to the process of urban growth. At that time, the establishment of *Parc-des-Plaines d'Abram* (1908) and the extension of *Boulevard Saint-Cyrille* (*Boulevard René-Levesque*) (1911) redirected the process of urban development in this direction. Indeed, it was in the beginning of the twentieth century that the privileged links were established with the *Boulevard Saint-Cyrille*, in detriment of the former expansion of the city in the direction of *Chemin Sainte-Foy*. The 1923 map shows this reorganization. At that time, *Avenue de Bourlamaque* was not connected with *Chemin Sainte-Foy*, ending at the intersection with *Rue Père Marquette*.

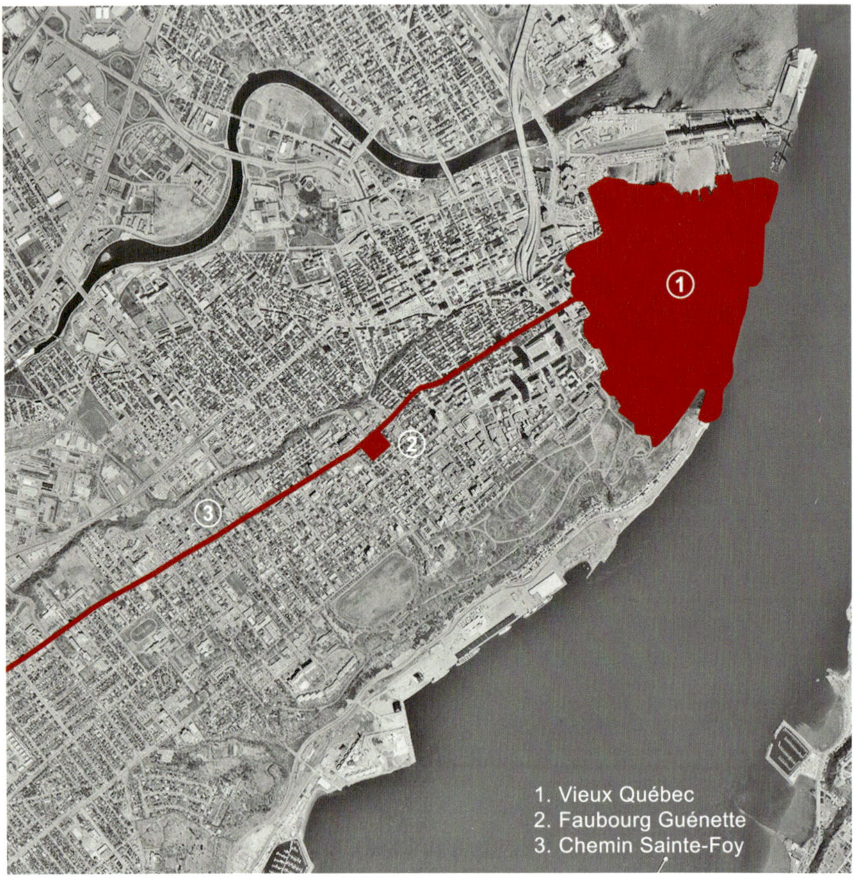

Fig. 9.1 The *Faubourg Guénette* site

It was in this stride, in 1912, that the building on our plot was erected. The main facade of the building and the entrances were oriented towards *Avenue de Bourlamaque*, an axis leading into the tramway and the Plains of Abraham. Along *Rue Crémazie*, the building facade precedes the backyard which remained free of construction until our intervention. At this time, two urban systems are still disconnected separated by *Rue Crémazie*. This division is still visible; even today, built types of different nature face each other along this street (Fig. 9.3).

Figure 9.4 shows the process of connecting *Avenue de Bourlamaque* with *Chemin Sainte-Foy* in the end of the 1920s, through the demolition of a number of suburban houses located north-west of our plot and *Rue Crémazie*. New larger buildings were then build, constituting a facade for this restructuring street.

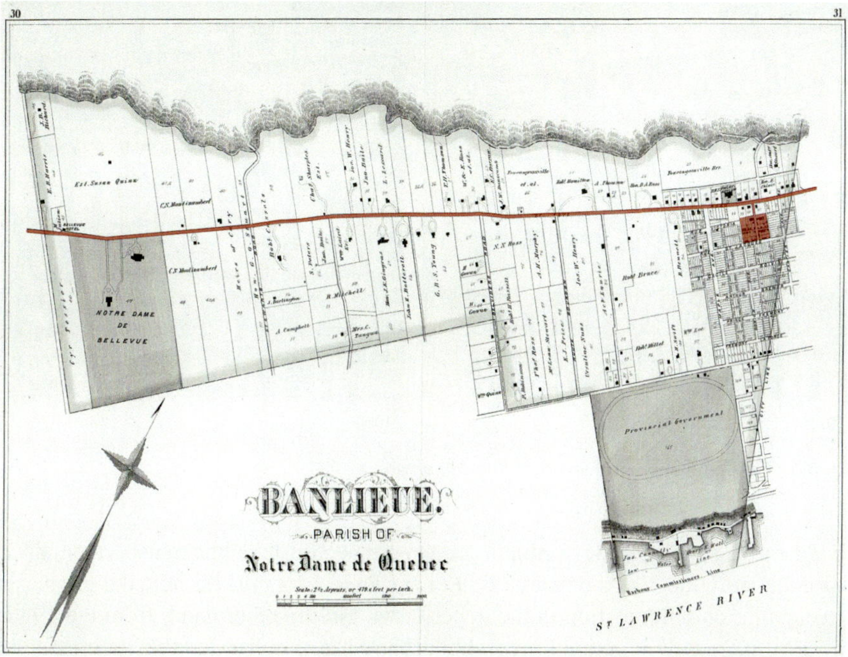

Fig. 9.2 Parish of *Notre Dame de Quebec*, by HW Hopkins, 1879 (*Source* Public Domain)

Fig. 9.3 The *Rue Crémazie* as a limit between two distinct morphological areas (*Source* Photograph by the author)

Fig. 9.4 The completion of *Avenue de Bourlamaque*

If at the beginning of the twentieth century connection to public transport remained essential, from the 1960s onwards *Boulevard René-Levesque* became the main axis of vehicular communication in the upper town. On the other hand, from the 1980s onwards, increased attention for neighbourhood life, emphasized the importance of *Rue Cartier* and, by extension, of *Rue Crémazie* as places lived at a pedestrian scale by the inhabitants of the neighbourhood, with its vibrant small shops and restaurants for a population looking for urban life.

At the time of the intervention, at the junction of *Rue Crémazie* and *Avenue de Bourlamaque*, half of our plot—in contact with *Crémazie*, was used for parking. The old building was divided into five units: four dwellings with access from *Avenue de Bourlamaque* and a small shop opening to *Rue Crémazie*. Over time, several parts have been added to the building towards the courtyard and poor-quality constructions were added between the duplex and the neighbouring building along *Avenue de Bourlamaque*.

The surrounding environment is heterogeneous. On the north side of *Rue Crémazie* there are the narrow streets of the former *Faubourg Guénette* and a number of small houses. At west of the plot, in *Avenue de Bourlamaque*, there is a twelve-stories social housing complex overlooking this entire street front. Finally, in the street block of our plot, there are mostly triplex buildings, common in the urbanization of this part of the city in the beginning of the twentieth century, built at the same time of the former building of our plot.

9.3 The Infill Strategies

The project targets the interstices created by a serial construction that has taken place over time in the plot. As pointed out by Caniggia and Maffei (1979) this situation is part of the spontaneous process of formation of the street block, in which a progressive densification takes place until the completion of a continuous built frontage in the perimeter of the street block (Fig. 9.5). As such, the project aims at highlighting the organic role of the intersection of *Avenue de Bourlamaque* and *Rue Crémazie*. The project is also part of a logic of progressive densification of the neighbourhood. At first, the elements added over time are demolished to make way for the new built parts which aim at the completion of the plot and urban continuity (Fig. 9.6).

A thin volume, three-storeys high, allows both to create a built frontage along *Rue Crémazie* and to give an identity to the new house with its main entrance. This also allows to continue the process of densification of the plot while contributing to

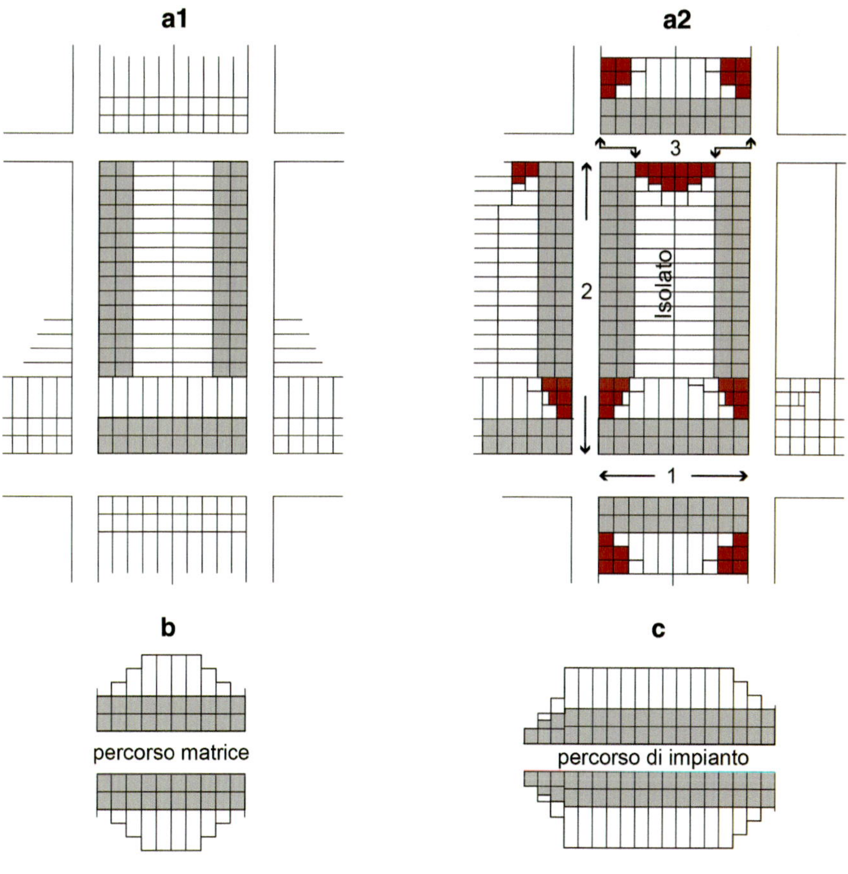

Fig. 9.5 The formation process of the street block (*Source* Caniggia and Maffei 1979)

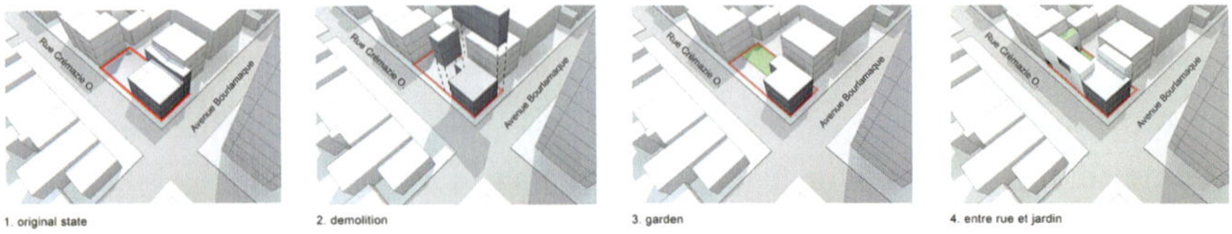

Fig. 9.6 Between the street and the garden—conceptual diagram

Fig. 9.7 View from *Avenue Bourlamaque* (*Source* Photograph by Félix Michaud)

the character of neighbourhood street constituted by the quiet and local living *Rue Crémazie*. On the other hand, the construction of a new volume in the gap left vacant in *Avenue de Bourlamaque* makes it possible to insert a new and thin three-storeys house (Fig. 9.7).

9.4 An Urban Project 'Between the Street and the Garden'

The project was born out of the desire of a family to settle in the heart of the urban and intense life of the *Montcalm* district in Quebec. After considering the feasibility of the project, it was decided to build a dwelling of around 300 m^2 on the open area of the plot and to redevelop the existing building into two different dwellings.

The owners aspired to live in a large open space; they were envisaging a loft, but the shape of the site, its dimensions and the layout of the duplex did not allow the development of such a housing type. Simultaneously, they also appreciate hidden corners and spaces. This duality inspires the configuration and the organization of the project. The house is spread over three levels of a long and narrow volume, located on a thin strip of land. It is conceived as an 'interval' architecture emerging in the urban fabric. This porous architecture threads its way between the street and the former 'void' of the plot. Inside, the house is enriched by the complexity of the site.

At the heart of the new volume, a generously glazed vertical core organizes the living rooms of the house, in the longitudinal axis: the living room and the kitchen in the ground floor; a music room, a dance studio and a small library in the first floor. In the transverse axis, the core concentrates the visual and physical links between the street and the garden, two highly contrasting environments. In the south side, the wall is pierced with large openings which give access to the garden of abundant vegetation (Fig. 9.8). In the north side, contact with the street is modulated by a perforated stainless-steel filter, superimposed on the bay windows: the screen. This element allows residents to perceive urban life while being protected from the look of pedestrians. The facade of the building translates the presence of the vertical core and divides the length of the new building into three portions, with similar proportions to the neighbouring triplex buildings (Figs. 9.8, 9.9, 9.10, and 9.11).

The pattern of holes in the 'fence', generated by a computer program, is designed to introduce a progressive variety in the arrangement, density and dimensions of the perforations. On the ground floor, there are fewer openings to preserve the privacy of the house. As they go up, they become more numerous, to favour natural light on the first floor, where it is less necessary to protect family and domestic life from outside looks. In the higher floor perforations continue to multiply and split into smaller modules, to ensure the safety of the railing of the *loggia*, an exterior extension of the house which oversees the vertical core.

While the new spaces built on the plot are very 'windowed' and directly connected to each other, the spatial structure of the duplex remains unchanged and hosts some of the 'hiding places' of the house. An opening barely wider than a door, in the first floor of the old exterior wall of the duplex, allows passage from the new to the existing building. In this axis, we can see the entire length of the house (more than 27 m), which corresponds exactly to that of the plot (Fig. 9.12). This 'threshold' marks the transition between very different atmospheres: from a large open and bright space into the muffled calm of three small rooms, each lit by one or two windows. In addition to the main frame and most of the partitions, other architectural elements of

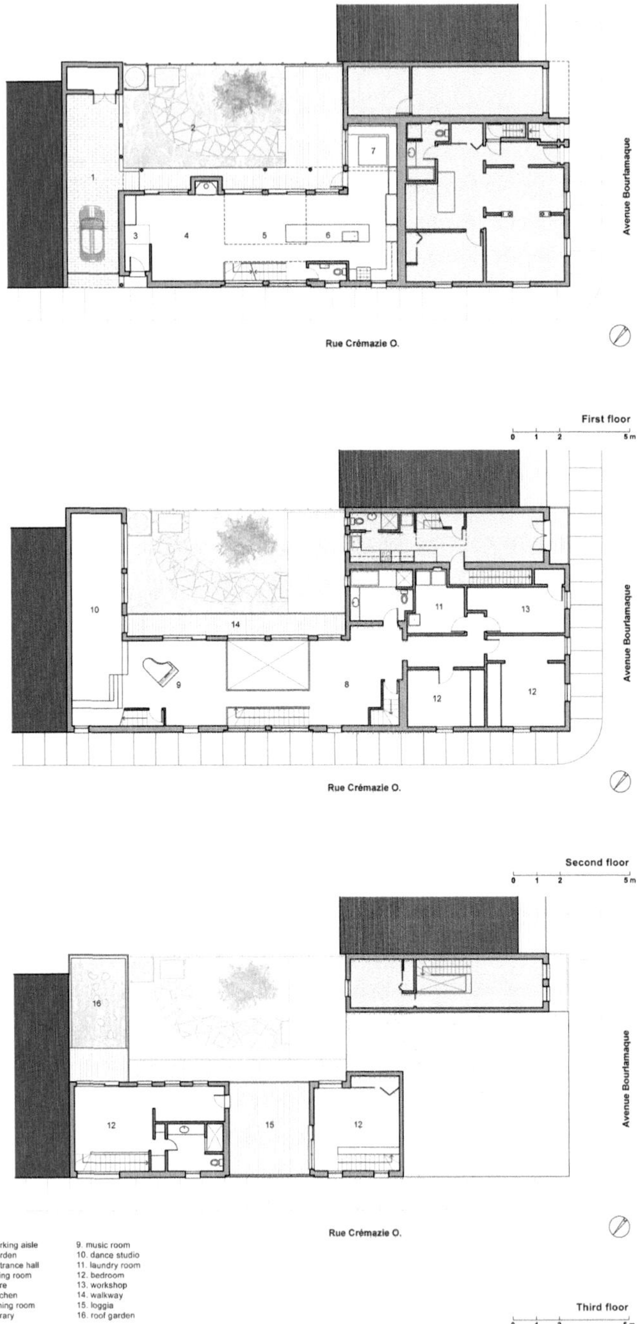

Fig. 9.8 Building plans

Fig. 9.9 Building sections

Fig. 9.10 View from *Rue Crémazie* (*Source* Photograph by Félix Michaud)

Fig. 9.11 The double height living room (*Source* Photograph by Félix Michaud)

Fig. 9.12 The full length of the new building (*Source* Photograph by Félix Michaud)

Fig. 9.13 The transverse *loggia* (*Source* Photograph by Félix Michaud)

the duplex have been preserved: certain doors, the mouldings and even, at the request of the owners' daughter, a green stain on a wall of wooden planks.

The last level of the house is the most 'secret'. On either side of the central core, two hidden stairs lead to two bedrooms. These rooms are shelters isolated from all domestic activity. They enjoy a magnificent view, over the roofs of the surrounding houses, towards north and the Laurentians mountains (Fig. 9.13). The two bedrooms give access to the *loggia*, a covered balcony, which one enjoys with pleasure in rainy, cool or sunny weather, day and evening. The *loggia* offers, outside and, at certain height, connections between the street and the garden, different from those of the lower levels. It also provides a very private exterior extension, despite the near presence of a large building.

9.5 A Small Garden in the Heart of *Montcalm* Neighbourhood

To compensate the reduced area of the plot and enrich the various spaces of the house, the exterior extensions are scattered through the different storeys. These spaces help to reduce the perception of density associated with life in an urban environment.

In the ground floor, the garden plays the central role of an exterior room and induces an expansion of the interior spaces (Fig. 9.14). With the different seasons, it defines an important part of the atmosphere of the living room and kitchen. The pale colour and reflective finish of the corrugated sheet that covers two of the three

Fig. 9.14 The garden as an extension of the living room (*Source* Photograph by Félix Michaud)

facades surrounding the garden transform these into light reflectors, for the benefit of the interior spaces and exterior extensions. Leaning against the fence, at the end of the stone path, an outdoor shower allows to cool off in hot weather and replaces the swimming pool of the former house of the owners.

In the first floor, the passage overlooks the garden and acts as a sun visor above the large bay windows facing south-east. Accessible from the music room, in line with the dance studio, it becomes a place of rest for a break during a rehearsal. It offers another point of view and contact with the garden. In the upper level, in addition to the *loggia*, a hanging garden extends the one on roof of the dance studio and offering a plant-wide landscape. Even if the dimensions of these outdoor spaces are reduced and they are not 'useful' in winter, they play an essential role in the atmospheres of the various places of the house. They contribute significantly to the pleasure and well-being of the occupants who do not miss their large suburban courtyard and do appreciate the qualities of their architectural and urban living environment (Fig. 9.15).

Fig. 9.15 The roof garden as an extension of the bedroom (*Source* Photograph by Félix Michaud)

9.6 An Architecture Where Domestic Spaces Meet Urban Life

The analysis of the formation process of the urban fabric of *Montcalm* district makes evident that beyond the first construction, several revisions have taken place in the urban organization. Originating on *Chemin Sainte-Foy*, the first development of the area was done at the edge of the countryside with the modest houses of the *Faubourg Guènette*. The urban form and the building typologies of these first interventions are still legible and contribute to the current character of *Rue Crémazie*.

The fabric was then reoriented following the introduction of the tramway and the Plains of Abraham Park. This had an important effect on the organization of the fabric itself as well as on the building types which were gradually introduced. At that time, the first construction on our plot is erected—a duplex oriented towards *Avenue de Bourlamaque*, in the direction of new urban facilities such as the tramway and Abraham Park.

The infill project follows major transformations in the renewed urban character of *Rue Crémazie* and, more particularly, in the friendly and commercial street character of *Rue Cartier*. This is a lifestyle that has attracted owners to value and inhabit an empty and neglected parking plot. Reading the formation process of tissue and building also highlighted how, over time, the organic role of the intersection of *Avenue de Bourlamaque* and *Rue Crèmazie* was formed.

The project proposes a porous architecture that combines the abstraction of the pure volume, with the complexity of its location. It is an architecture which dialogues both with the historical stratification of the site and the life which has gradually taken place in the neighbourhood. As a 'medium', this porous architecture creates a link between the privacy of domestic life and the richness of the urban space. The wall, the usual delimiter of liveable space, becomes itself inhabited and free from its definition as a 'limit', merging at the same time with the public space of the street and the privacy of the garden. Those who inhabit it find themselves in direct contact with the surrounding urban landscape, in the hollow of an interval-architecture.

The project responds both to the process of transformation of the urban fabric and to the new stages and meanings of the community. The proposed architecture allows individuals to find 'opportunities' to fulfil their aspirations in the general context of the urban edification process. More directly, the changing context of the North American city makes it possible to grasp this process of accumulation of progressive meanings, dialogue between society and individuals, which gradually constitutes a built framework, re-interpretable and, for that reason, increasingly rich in its complexity.

References

Caniggia G, Maffei GL (1979) Composizione architettonica e tipologia edilizia I: lettura dell'edilizia di base. Marsilio, Venice

Cervellati PL, Scannavini R (1975) Bologna, politica e metodologia del restauro nei centri storici. Il Mulino, Bologna

Dufaux F (2007) The origins of Montreal' s housing tradition. Unpublished PhD thesis, University College London, United Kingdom

Muratori S (1959) Studi per una operante storia urbana di Venezia. Istituto Poligrafico dello Stato, Roma

Muratori S, Bollati R, Bollati S, Marinucci G (1963) Studi per una operante storia urbana di Roma. Consiglio Nazionale delle Ricerche, Roma

Rossi A (1966) L'architettura della città. Marsilio, Padova

Rossi A (1976) La città analoga. Lotus. International 13:5–9

Chapter 10
On a Morphologically-Based Method for Architectural Practice: The *Lindo Vale* House

Vítor Oliveira and Cláudia Monteiro

Abstract This chapter is in two parts. The first part proposes a morphological reading grounded on the town-plan concept. The concept is applied in the analysis of *Lindo Vale*, a nineteenth-century street in the city of Porto, Portugal, that has suffered profound transformations in the last decades of the twentieth century. The second part of the chapter describes how this morphological approach informs the design of a new house in *Lindo Vale*. It addresses explicitly one fundamental question of the relation between analysis and design of the urban landscape: what to conserve and what to change? This architectural project conserves the most structural elements of *Lindo Vale* street and its urban landscape: the plot width, the coincidence between plot frontage and building frontage, the dominant relation between the height of the new building and the width of the street, the overall interior organization of rooms and the position of the staircase within the building. The project introduces a new architectural style, a new façade design and new construction materials. The specific nature of this building as a design answer to a particular urban landscape is highlighted through a comparison with two other buildings designed by its authors in the city of Porto—one in a medieval street, the other in a street built in the first half of the twentieth century.

Keywords Research and practice · Urban morphology · Urban form · Architecture · Conservation and change

V. Oliveira (✉) · C. Monteiro
Universidade do Porto, Porto, Portugal
e-mail: vitorm@fe.up.pt

C. Monteiro
e-mail: anaclaudiapmonteiro@gmail.com

V. Oliveira (ed.), *Morphological Research in Planning, Urban Design and Architecture*, The Urban Book Series, https://doi.org/10.1007/978-3-030-66460-2_10

10.1 A Morphological View of Cities and City-Building Processes

Humankind started living in cities about 6000 years ago (Morris 1972; Schoenauer 1981). This long period of urban life was made of permanencies and ruptures in a learning process of continuous adaptation of urban forms—streets and squares, street blocks, plots, special and ordinary buildings, to name the most important—to human needs and aspirations. When looking at the 'big picture', at this long time period of six millennia, it is evident that the major rupture happened 'yesterday', or to be more precise in the last century. While some expansions of the mid-nineteenth century, started to change the nature of street blocks increasing their role as 'forms' and decreasing their importance as 'processes' (gathering the action of different agents), it was in the early twentieth century that the major changes happened, when buildings started to progressively setback from streets (Castex et al. 1977; Dibble et al. 2017; Oliveira et al. 2020). By the mid-twentieth century mainstream architecture was exclusively focused on buildings as autonomous objects in a new urban landscape where the street had lost its role, and mainstream planning (and traffic engineering) was focused on streets as the privileged support for vehicular movement and on land and building utilization as the main component of zoning maps.

Criticism to these new processes of city building arose as soon as their results on the ground, and on the everyday life of people, become evident. From the end of the 1950s and early 1960s to the end of the 1970s major research works have been prepared, describing the problem, explaining its causes and offering evidence for prescription. Despite their apparent significance on academia, their impact on practice has been reduced. Today, six decades after the publication of 'Studi per una operante storia urbana di Venezia' and 'Alnwick Northumberland: a study in town-plan analysis', and four decades after the writing of 'The social logic of space', the major weaknesses and threats characterized in these seminal books, and by the morphological schools of thought they have inspired, remain or have been amplified (Conzen MRG 1960; Hillier and Hanson 1984; Muratori 1959).

How to recover the physical relation between past, present and future? It is argued that the town-plan, as a unifying concept, is capable of relating these different time periods and, as such, of establishing a link between research and practice (Oliveira 2020). The town-plan concept has been originally proposed by MRG Conzen as part of the tripartite division of the urban landscape (Conzen MRG 1960). It is made of streets (including squares and gardens), plots and the block-plans of buildings. The other two elements of the tripartite division are the building fabric and the land and building utilization. The potential of the concept is fully illustrated in the seminal book of Alnwick, which despite Conzen's original intentions of embracing the three elements has remained a study on town-plan analysis (Monteiro 2017). The concept offers the stage for the creation and development of other concepts that address the way urban form elements are combined on the ground and how they evolve over time, notably the morphological region (Conzen MRG 1975; Whitehand 2009; Oliveira and Yaygin 2020), the fringe belt (Whitehand 1967, 1972, 2019; Conzen MP 2009; Ünlü 2013; Oliveira 2019) and the morphological period (Conzen MRG 1988). Due to their conceptual complementary nature, the morphological period can be related to a concept with a high potential for relating analysis and design, the typological process (Caniggia and Maffei 1979; Monteiro and Pinho 2021; Whitehand et al. 2014). Finally, Conzen MP (2018) offers a notable synthesis on town-plan analysis as a method for understanding the physical evolution and the present character of cities.

In this chapter, it is argued that the town-plan elements that can be used in the description and explanation of different areas can also be utilized in the design of their future transformation. The way the system of streets, squares and gardens is organized in a city, as well as the density of its elements and its intersections, allowing more or less spatial accessibility, and thus favouring or hindering the flows of movement of its residents, workers and visitors, is a decisive factor in structuring a territory and in promoting effective urban cohesion. Each transformation of this system, given its high permanence in time (it is the most permanent of urban form elements), must be thoroughly considered. The definition of a street system is always associated with the conformation of a block system. In a way, the two systems correspond to the 'full' and 'empty' part of the same object. The first system guarantees urban flows, and the second provides support for the definition of plots and construction of building stocks. In each street block, the density of plots is somehow related to the presence of agents and, as such, to the diversity of urban strategies. Despite its 'invisible' nature, the increase in plots size and the reduction of the number of agents has been one of the most significant changes that occurred in the twentieth century in the physical form of cities, with consequent losses in different aspects of urban life. Each building to be erected on each of these plots will confirm its diversity potential. It should also actively contribute to the formal definition of the street as an attractive place for different modes of transport, particularly the pedestrian mode.

10.2 The Morphological Analysis of *Rua do Lindo Vale*

Rua do Lindo Vale is an early nineteenth-century street, that was part of one of the five gateway paths leading Porto to north, to the city of Guimarães. Figure 10.1 gathers four of the 400 sheets of the notable 1892 map, the first to represent the city as it is known today and, based on that, Fig. 10.2 offers a reading of streets, plots and block-plans of buildings. *Lindo Vale* is located two kilometres north of Porto's fourteenth-century wall. In mid-nineteenth century a new street (*Costa Cabral*) has been built, east of the former. Being longer and wider, *Costa Cabral* soon took a more important role in the street system of Porto. *Lindo Vale* assumed a residential character. *Lindo Vale* is 500 m long and it is made of two segments (Fig. 10.2). The first segment, located south, between *Marquês do Pombal* square and *Álvaro de Castelões* street, is about 300 m long and 6–10 m width (it has a slight change of direction and width). The second segment, located north, between *Álvaro de Castelões* and *Cunha* streets,

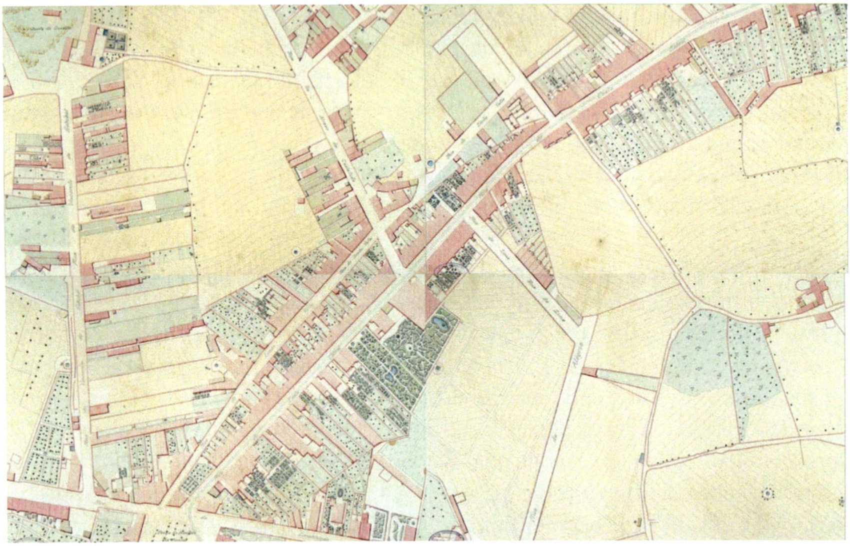

Fig. 10.1 *Rua do Lindo Vale* in the 1892 map by Telles Ferreira—four sheets at 1:500 scale survey (*Source* Public domain)

is about 200 m long and 10 m width. The 1892 map (and its reading in Fig. 10.2) makes evident that the second segment was, in the end of the nineteenth century, in an earlier process of urban development, with many large and empty plots.

Lindo Vale street is shaped, at east, by two small (around 10.000 m^2) elongated street blocks and, at west, by two street blocks of undefined form (including rural paths). The plot structures of the four street blocks have some differences. The east side of the first segment is made of 42 plots. Most plots are narrow, 5–6 m frontage (see Table 10.1). Part of these plots is in contact with one street only—*Lindo Vale*, while others are also connected to *Costa Cabral* street. In the latter group of plots, building and plot frontages are coincident in *Costa Cabral* while their backyards face *Lindo Vale*. The west side of the first segment is made of 40 plots. Three major directions shape this set of plots. Most plots are narrow, 5–6 m frontage, with variable depth (plot depth of the west side is higher than of the east side). Building and plot frontages are coincident. Contrarily to the first segment, the second segment of *Lindo Vale* is made of both small and very large plots. The number of buildings in this segment is still reduced, particularly in the west side of the street, with some irregular patterns of occupation. The east side of this second segment is made of eighteen plots, and the west side is made of twelve plots.

Figure 10.3 gathers a set of photographs of some of the 70 original buildings of *Lindo Vale* (nineteenth and early twentieth century) that have survived until the twenty-first century. In a way, it offers a simplified view of the typological process (as a succession of types in the same cultural area) that has occurred in *Lindo Vale*—Fig. 10.4. It has been a continuous evolution from narrow-frontage plots (5–6 m)

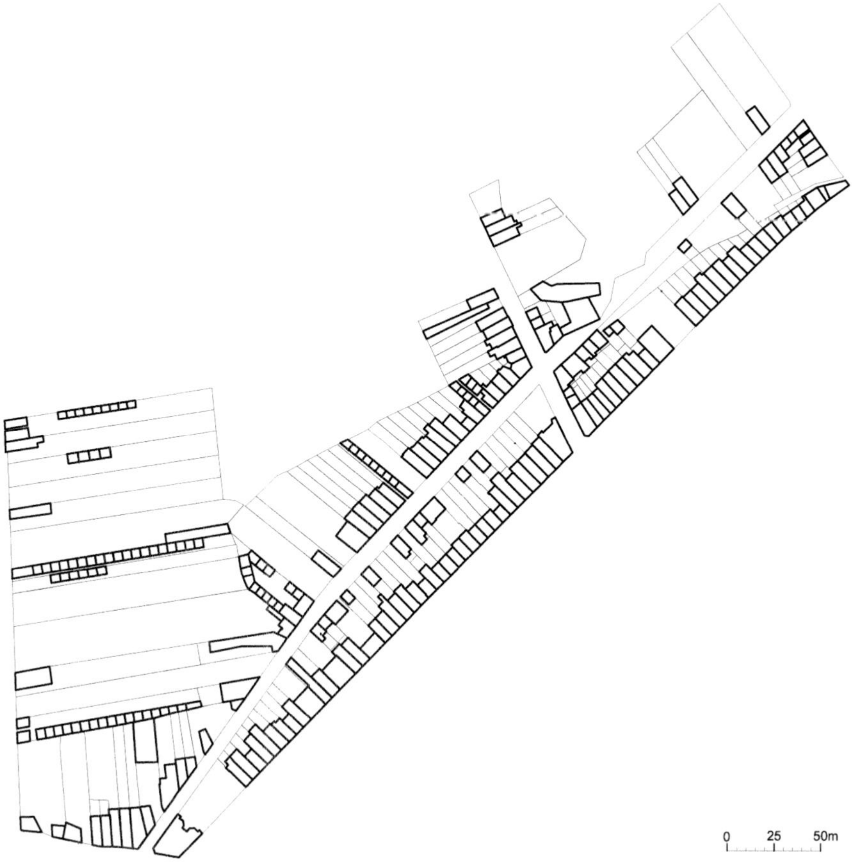

Fig. 10.2 *Rua do Lindo Vale* in 1892: streets, plots and block-plans of buildings

Table 10.1 Plots and buildings of *Rua do Lindo Vale*: 1892, 1960 and 2020

	Segment 1				Segment 2				Total	
	East		West		East		West			
	P	B	P	B	P	B	P	B	P	B
Late nineteenth century, 1892	42	14	40	37	18	13	12	8	112	72
Mid-twentieth century, 1960	47	27	47	41	30	24	30	29	154	121
Early twenty-first century, 2020	48	39	43	38	24	22	26	26	141	125

P—Plots. B—Buildings

Fig. 10.3 Original buildings of *Rua do Lindo Vale*: increasing plot width (left to right) and increasing number of storeys (up down) (*Source* Photographs by the authors)

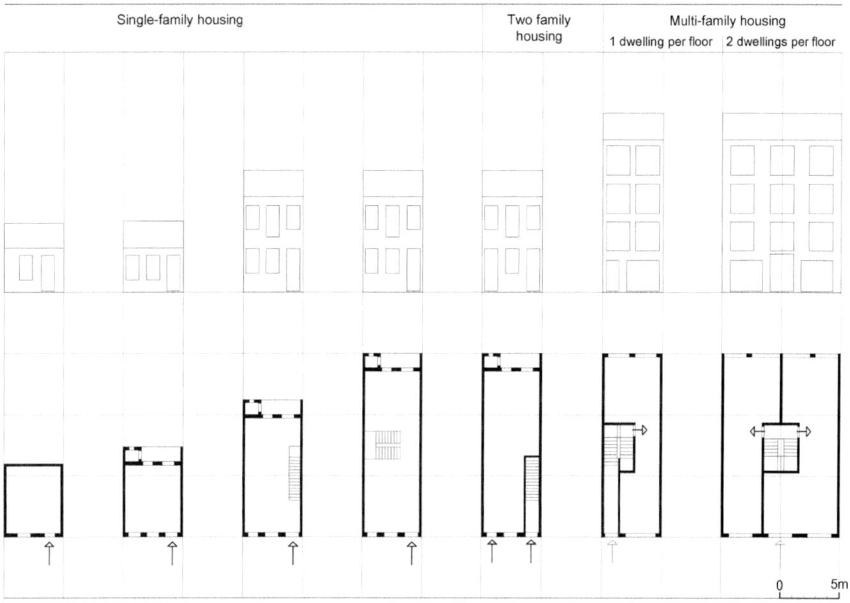

Fig. 10.4 A simplified typological process of *Lindo Vale'* buildings

and one-storey buildings with three bays (there is not in *Lindo Vale* a one-storey building with two bays, as it still exists today in *Costa Cabral*—Oliveira et al. 2015), to medium-width plots (13 m) and three-storey buildings with five bays. The four street-blocks include also a number of *ilhas*, a working-class housing type consisting of a bourgeois house in the plot head and a number of small houses in the plot tail, in a process very similar to that described by MRG Conzen (1960) as the burgage cycle.

Figure 10.5 and Table 10.1 make evident the town-plan transformations that have occurred in this area over seven decades, from 1892 to 1960. These changes are most evident in the second segment of *Lindo Vale*. Here, the number of plots increases from 18 to 30 in the east side, and from 12 to 30 in the west side. The large plots that existed in the end of the nineteenth century have all been subdivided, except the plot of *Júlio Deniz* cinema (offering its back, and one small door, to *Lindo Vale*) with 35 m long and 23 m width. In the first segment of *Lindo Vale*, the number of plots increases from 42 to 47 in the east side (with a very regular plot series) and from 40 to 47 in the west side. Most of the 154 plots and 121 buildings that existed in the mid-twentieth century had narrow frontages.

While between the late nineteenth century and the mid-twentieth century there have been many plot subdivisions, between mid-twentieth and early twenty-first century there have been ten processes of plot amalgamation. These processes usually involved the amalgamation of two plots, but in two cases it corresponded to the merging of three and four plots (Fig. 10.6). The total number of plots decreases

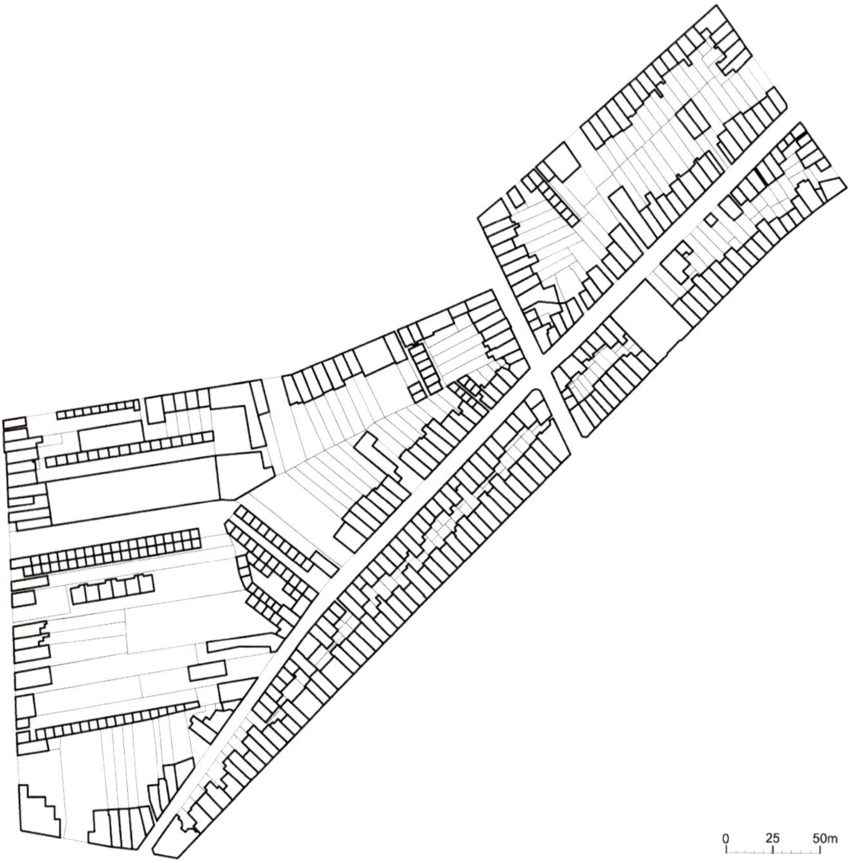

Fig. 10.5 *Rua do Lindo Vale* in 1960: streets, plots and block-plans of buildings

Fig. 10.6 Plot amalgamations in *Rua do Lindo Vale*: two, three and four plots (*Source* Photographs by the authors)

Fig. 10.7 *Rua do Lindo Vale*, aerial view (*Source* Google Earth)

from 154 to 141, between 1960 and 2020 (Figs. 10.7 to 10.9). The total number of buildings had a slight increase. Plot width in 2020 varies from 5 to 23 m, while plot depth goes from 7 to 56 m. Building height varies from one to four storeys, making the street section quite diverse. In most cases, building and plot frontages are coincident. While it is mainly a residential street, *Lindo Vale* gathers some other uses: retailing, services and warehousing. In addition to the most profound changes, involving plot amalgamation, there was also the demolition of extant single-family buildings for the construction of multi-family buildings on the same narrow plots. On the contrary, in the last two decades, building applications have led to re-building or the construction of new buildings for single-family housing (Fig. 10.8).

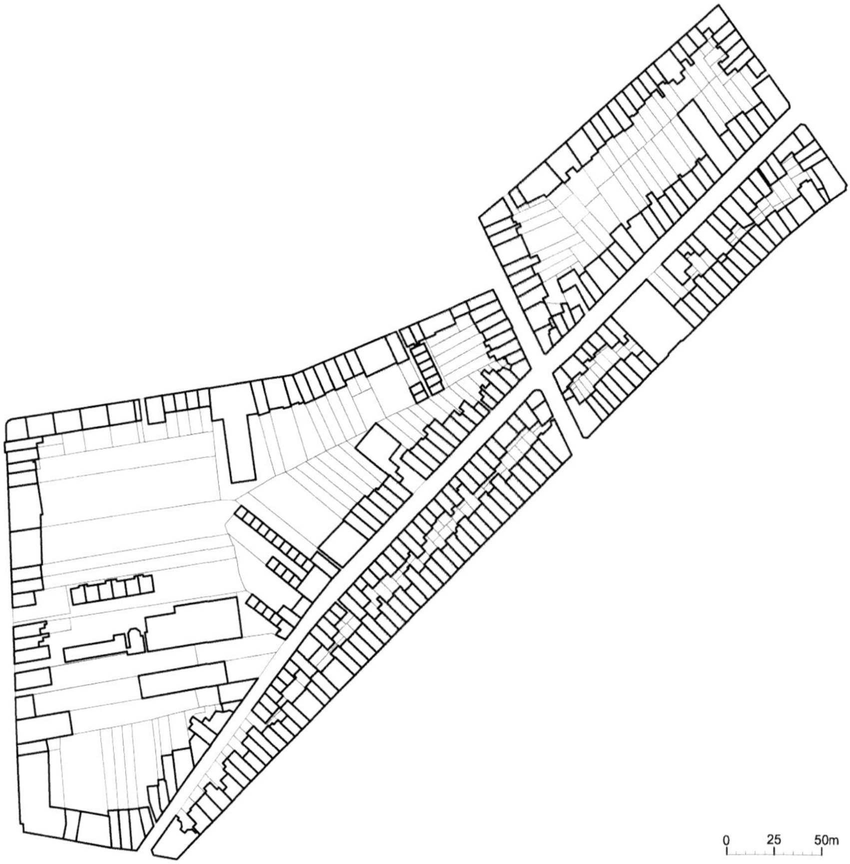

Fig. 10.8 *Rua do Lindo Vale* in 2020: streets, plots and block-plans of buildings

10.3 The Design of a New Building in *Rua do Lindo Vale*

10.3.1 Conservation

The plot for the new building has a regular shape, near to a rectangle, and its natural support is almost flat (*Rua do Lindo Vale* 435, 41°09′54.03″ N 8°36′03.71″W). The plot is 5.5 m wide and 40 m long (220 m^2 area). It is part of a series including seventeen plots of similar width and depth located in the west side of segment two, as shown in the 1960 map, in Fig. 10.5. While this plot series results from the subdivision of some large plots, represented in the 1892 map, it would suffer some processes of amalgamation in the second half of the twentieth century, as can be inferred by the comparison of the 1960 and 2020 maps. Where plot amalgamation has occurred

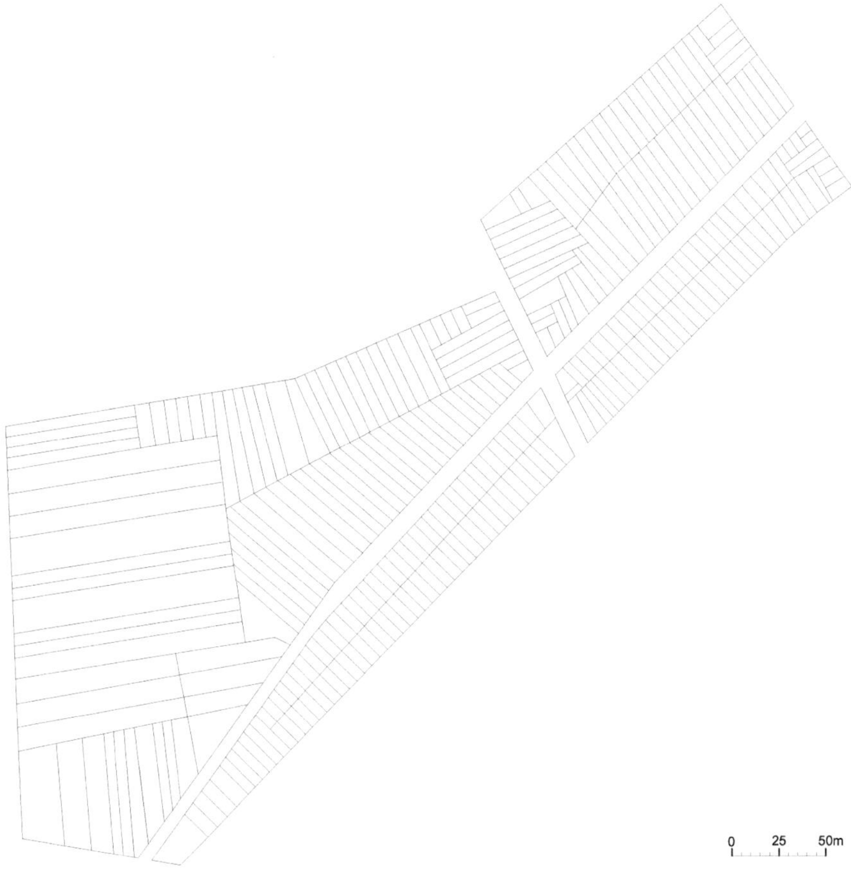

Fig. 10.9 *Rua do Lindo Vale*—hypothetical metrics of plots

in Porto, in recent decades, it has usually preceded the construction of apartment buildings, not of single-family houses.

The old building, that previously stood on the plot, with one-storey and 65 m^2 area, had already been demolished when the plot was acquired; only a part of the building's main façade remained intact. Reconstruction of the building and extension into a three-storey building (the only possibility permitted by the plan in force for designing a house of about 200 m^2) was not a feasible option, mainly because of technical and financial reasons.

The influence of the municipal plan on the architectural project will not be expanded in this chapter as it has already been explored by Oliveira and Monteiro (2014, 2016a, b) making evident the coincidence between the view of the plan and of this morphological analysis.

The new building contributes to maintenance of the existing relationship between buildings and street (Fig. 10.10). It follows both the front and rear alignments of

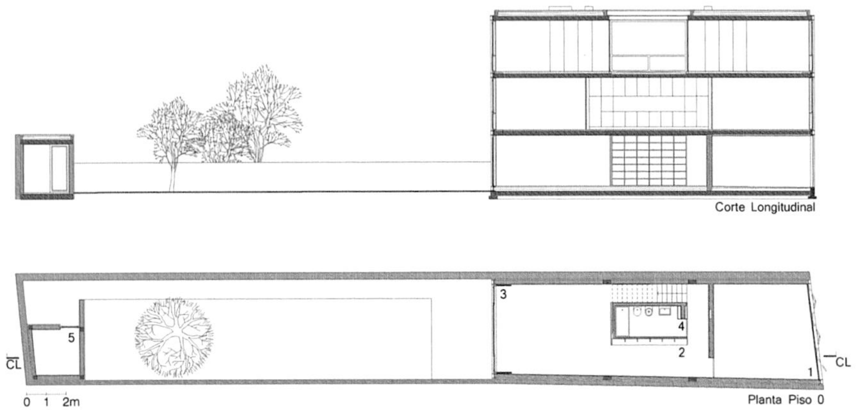

Fig. 10.10 House—section and ground floor plan: **1** patio/garage, **2** hall/library, **3** office, **4** bathroom, **5** laundry

buildings in this segment of the street. The building is three storeys and its height is 8.8 m (1.2 m less than the width of the street). The rear of the plot is a garden of about 21 × 5 m; its area is higher than the area of the building—building coverage is 46% of the plot area (Fig. 10.11). Although it is not the focus of this paper, this gives an idea of the efficiency of land utilization of these narrow long plots when compared, for instance, with plots of detached and semi-detached houses (usually larger and shallow). It is also important to mention that this is not a house built in the periphery of the city, in a plot of medium- or large-size (with their specific issues) but a house erected in the central area of the city, in an area with an intense urban life, in a type of plot that raises a number of spatial constraints.

Conformity to the size of existing buildings requires an interior arrangement of rooms similar to that of surrounding buildings, and of traditional houses built in Porto until the first decades of the twentieth century (Fig. 10.10). This means the organization in two lit rooms near the two façades of the house and a third non lit room in the core of the house (Fig. 10.11). This type of general organization, as in most buildings erected in *Lindo Vale* in the nineteenth and early twentieth centuries, leads to a location of the staircase—a fundamental element in the spatial interior organization—in the core of the house. The two rooms located near the two façades are: in the ground floor, a patio/garage and an office; in the first floor, one dining and one living room; in the second floor, two bedrooms. The third room, in the core of the house, corresponds to: in the ground floor, the library (which is opened to the office) and the bathroom; in the first floor, the kitchen (which is opened to the dining and living rooms); and, in the second floor, a patio (Fig. 10.12).

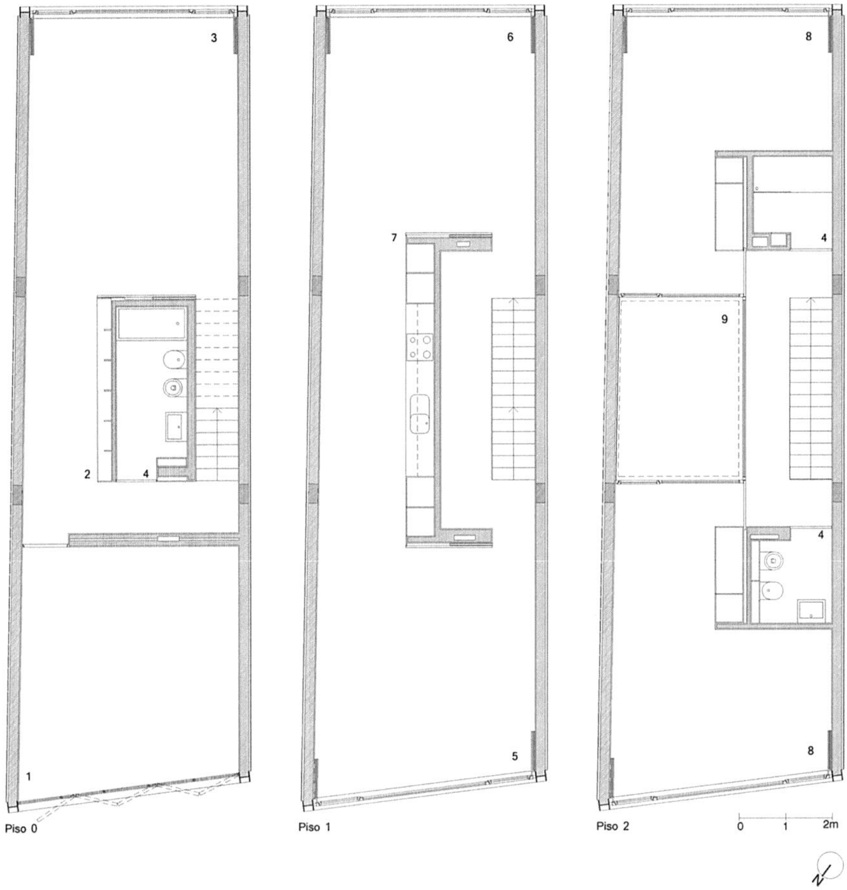

Fig. 10.11 House—plans of the different floors: **1** patio /garage, **2** hall /library, **3** office, **4** bathroom, **5** dining room, **6** living room, **7** kitchen, **8** bedroom, **9** patio

10.3.2 Change

What elements of change have been introduced in the street by this house? One of the most significant elements of change has been the architectural style (Fig. 10.12 and Table 10.2). While preserving a number of elements of urban form (including the definition of a building volume that is very close to that of existing houses), a new building language distinct from the buildings erected in the nineteenth century and in early twentieth century, and somehow close to modernist architecture, has been introduced. The house has a symmetrical and classically proportioned design, contains similarly sized elements, and has a neutral range of colours and materials. This option is grounded on the acknowledgement that the character of this urban landscape depends, not on the conservation of the existing buildings (or the construction

Fig. 10.12 The core of the house: library (ground floor), kitchen (first floor) and patio (second floor) (*Source* Photographs by José Campos)

Table 10.2 The house and the street—conservation and change

	Conservation	Change
Street	Relation between building height and street width	–
Plot	Plot width Coincidence between plot and building frontages	–
Building	Overall interior organization of rooms Position of the staircase within the building	New architectural style New façade design New construction materials

of new buildings of a similar architectural style), but of the existing plot structure and the relation between building and street.

This stylistic option raises a fundamental issue. The modernist movement has provided the needed instruments to establish a rupture with the way cities were built until the beginning of the twentieth century. It is argued in this chapter that the process of city-building with a focus on the 'building' and not on the 'street' has introduced a number of problems in our cities—physical problems, in a first moment, but also social, economic and environmental problems, in latter moments. This leads to question (and that could be the theme of a whole new chapter) if a return to a traditional way of making cities, with a focus on the street (and not on the building) as a fundamental urban element, could include a modernist architectural repertoire (and not a modernist planning repertoire).

One of the main characteristics of this house is the simplicity of design and of the new materials. The integration of the house within the existing buildings is based, in a first moment, on a careful volumetric articulation and, second, on the simplicity of design of the front and rear façades—the steel structure, the wooden door and windows, and the steel gate—reinforced by the use of one single colour. Inside the house this simplicity continues: in an interior organization layout that is very similar in all floors; in the search for 'truth' in construction (without 'false' ceilings or other elements that might be used for a spatial redefinition after construction); and in a spatial continuity between the three floors, which is enhanced by the design of the stair and of the vertical movement. This simplicity is also underlined by the use of a restricted range of materials in all rooms, including both the kitchen and the bathrooms: walls and ceilings are painted white, pavements are of beige vinyl floor, and wooden interior sliding doors and shutters are painted, or lacquered, in white.

Another important characteristic is the flexibility of the house. The repeated use of a set of very simple elements introduced a great versatility in the day-to-day living of the house. Firstly, each window, on the front and rear façades, has two sets of shutters, allowing the establishment of different ambiences in each room. Secondly, all rooms, except the bathrooms, have sliding doors that can be fully collected by the interior walls allowing, as such, a smoothness movement between different rooms, or the isolation of these rooms. Other elements—the garden, the patio in the second floor or the flat roof window over the shower in one of the bathrooms, to name just a few—offer a sound contribution to the spatial qualification of the house. Finally, there is also flexibility of uses. Being the home and the working place for two architects that spend considerable time at office, the choice of its location within the building is crucial in terms of energy consumption. Inspired by traditional Islamic architecture and, to a certain extent, by some Portuguese rural houses, the office in *Lindo Vale* has a dual location. It is in the ground floor in the hottest months of the year, between early May and late October, and in the east room of the second floor in the coldest months of the year, from early November to late April.

The design of the architect's house is, undoubtedly, a unique challenge within the design of the 'house', as an architectural programme, due to the coincidence between the 'architect' and the 'client', and due to the possibility of the spatial realization of the architect's own residential experience. The house in the *Rua do Lindo Vale* is the spatial expression of the residential experience and of a way of living of their architects; erected on a great simplicity of design and materials, achieving its high quality through a rigorous sense of construction and a high spatial versatility. But, most of all, it is clearly an urban house belonging to 'that' particular place. This is explored in the next section through a brief comparison of this house with two other architectural projects designed by the same architects for different urban landscapes in Porto (Figs. 10.13 and 10.14).

Fig. 10.13 House—front and rear façades (*Source* Photographs by José Campos)

10.3.3 Conservation and Change in Different Urban Landscapes

The first building is in *Rua do Loureiro*, a medieval street, south of the central train station, once contained within the fourteenth-century wall (Fig. 10.12). The relief of this area has significant variations, going from 55 to 70 m. The street is small (about 150 m long), narrow (4–8 m width) and irregular. Plots are also narrow and irregular. There is high building density. Buildings are narrow, normally four storeys high (some have three or five storeys). Although buildings are always positioned at the front of the plot, building coverage is very high.

The building is located in the north-south segment of the street (the shortest of the two segments that make it), in its east side (*Rua do Loureiro* 67, 41º08′40.72″N 8º36′34.11″W). Despite the medieval origin of the street, the building has been erected in the 1930s, with an architectural style close to that of the beginning of the twentieth century. The architectural programme, developed in the last decade, involved an intervention in the whole building (a multi-family building with commerce in the ground floor), adapting it for local accommodation. Due to the urban history of the place, the physical characteristics of the street and of the building, the option was to conserve as much as possible: the façade, including its doors and windows; the position of the staircase; the interior arrangement of rooms; and the materials and design in the interior of each apartment. The building, both in the exterior and in the interior (and this is a crucial aspect), is a building of the first half of the twentieth century (Fig. 10.15).

The second building is located in *Rua Domingos Sequeira*, a street that has been built in the first half of the twentieth century. It is part of a residential neighbourhood north of *Boavista*, the main service centre of Porto. The street is 13 m wide, plus the additional setback of buildings (4–5 m). Plots have different shapes and sizes (usually 7–15 m width, and 25–40 m depth). The street is made of detached and semi-detached buildings, two- to three-storeys, mostly single-family buildings.

The building, erected in the 1940s, is in *Domingos Sequeira* 154 (41º09′49.13″N 8º37′46.05″W). The architectural programme corresponded to an intervention in this single-family building with a focus on the basement floor, which in this type of building was usually used for garage and a number of very small rooms for ‘minor’ uses. The architectural project proposes conservation of building façades, position of the staircases, interior arrangement of rooms and materials in the first and second floors (Figs. 10.15c and 10.15d). On the other hand, it proposes transformation of the interior arrangement of rooms in the basement, creating a large room (with bathroom) and an office, related with a patio.

These three projects designed by the same architects for the same city (Porto) within a short period of time (five years), illustrate one of the main arguments of this section. The house of *Lindo Vale* is a building that belongs to ‘that’ street; a street that has been in constant transformation and where the main elements to conserve are the plot structure, the position of the building within the plot, and the relation between the building and the street. In the project for *Domingos Sequeira*, due to the

Fig. 10.14 *Rua do Loureiro* (medieval), *Rua Lindo Vale* (nineteenth century) and *Rua Domingos Sequeira* (first half of the twentieth century), approximately at the same scale (*Source* Google Earth)

Fig. 10.15 *Rua do Loureiro*, *Rua Domingos Sequeira* and *Rua Lindo Vale* (*Source* Photographs by the authors—**a**, **b**, **c**, **d**—and José Campos—**e**, **f**)

characteristics of the street and of the extant building, the architects were much more conservative, preserving the exterior and most of the interior of the house, except for the basement floor. Finally, when acting on a medieval street, the objective was to maintain as much as possible the extant physical elements and, as such, the identity and character of the building.

10.4 Conclusions

This chapter illustrates how the morphological reading (influenced by the historico-geographical approach and, in particular, by the town-plan concept) of a street can inform the design of a new building for that street. This process is not linear or straightforward. One particular morphological approach does not lead to one specific architectural design. On the contrary, what has been presented here is a method to describe and explain a particular urban landscape, to gain specific morphological knowledge on that landscape and, then, to apply it into an action of change that should be as adequate as possible to the particular characteristics of the urban landscape.

This process, from research to practice, focuses on the most permanent and structural aspects of urban form and not on non-essential characteristics. It is argued that cities are in constant transformation; that transformation is needed if seen through the lens of constant adaptation of urban forms to human needs. Each part of the city, each urban landscape, has its own rhythms of change and some of its elements are more susceptible to transformation than others. The selective conservation of some elements of urban form does not mean refraining this natural process of change or preventing architectural creativity. On the contrary, it means achieving better understanding of the process of transformation and establishing a framework for the action of the architect, making it less susceptible to the fashionable nature of the architectural system. Indeed, in times when architectural research, teaching and practice seem to converge into the production of spectacular isolated buildings, it would be most welcome the establishment of a new paradigm where architectural quality would be assessed by how well it relates to the particular urban landscape where it fits, and how well it addresses the needs of people who will use it.

Assuming the non-linearity of moving from research to practice (for different perspectives on this topic see, for instance, Holanda 2018 and Strappa 2018), in the end of the morphological analysis it should be clearer than in the beginning, what to conserve and what to change. This is a fundamental question in a morphologically-based design. The simplified morphological analysis of *Lindo Vale*—able to fit into an architectural practice with no additional resources needed—has revealed the most structural elements that should be conserved. These are the plot width, the coincidence between plot frontage and building frontage, the dominant relation between the height of the new building and the width of the street, the overall interior organization of rooms and the position of the staircase within the building. At the same time, it has suggested the elements that could be changed: the possibility of exploring a new architectural style, a new façade design and of introducing new construction materials.

The brief comparison of the house in *Lindo Vale* with the buildings in *Loureiro* and *Domingos Sequeira* has illustrated another important argument of this chapter—each specific urban landscape 'demands' a particular project. The steel and glass house of *Lindo Vale* would not make sense in the particular landscape of *Domingos Sequeira* and even less sense in *Loureiro*' s historical street, as the elements of change of the house would not fit into the physical characteristics of these two places.

Despite the difficulties of bridging the gap between morphological research and architectural practice, this chapter has made evident how a morphological method can influence the design of a project and produce a house that is in close dialogue with the street where it fits in.

References

Caniggia G, Maffei GL (1979) Composizione architettonica e tipologia edilizia I: lettura dell'edilizia di base. Marsilio, Venice
Castex J, Depaule JC, Panerai P (1977) Formes urbaines: de l'îlot à la barre. Dunod, Paris
Conzen MP (2009) How cities internalize their former urban fringes: a cross-cultural comparison. Urban Morphol 13:29–54.
Conzen MP (2018) Core concepts in town-plan analysis. In: Oliveira V (ed) Teaching urban morphology. Springer, Cham, pp 122–143
Conzen MRG (1960) Alnwick Northumberland: a study in town-plan analysis. Institute of British Geographers Publication 27, George Philip, London
Conzen MRG (1975) Geography and townscape conservation. In: Uhlig H, Lienau C (eds) Anglo-German symposium in applied geography, Giessen-Würzburg-München, Giessener Geographische Schriften, pp 95–102
Conzen MRG (1988) Morphogenesis, morphological regions and secular human agency in the historic townscape. In: Denecke D, Shaw G (eds) Urban historical geography. Cambridge University Press, Cambridge, pp 253–272
Dibble J, Prelorendjos A, Romice O, Zanella M, Strano E, Pagel M, Porta S (2017) On the origins of spaces: morphometrics foundations of urban form evolution. Environ Plann B: Urban Anal City Sci 46:707–730
Hillier B, Hanson J (1984) The social logic of space. Cambridge University Press, Cambridge
Holanda F (2018) Inserting urbanity in a modern environment. In: Oliveira V (ed) Teaching urban morphology. Springer, Cham, pp 185–203
Monteiro C (2017) Publishing and urban morphological classic. Urban Morphol 21:181–182
Monteiro C, Pinho P (2021) MAP: a methodology for morphological analysis and prescription. Urban Morphol 25:57–75
Morris AEJ (1972) History of urban form: before the industrial revolution. George Godwin Limited, London
Muratori S (1959) Studi per una operante storia urbana di Venezia I. Palladio pp 3–4
Oliveira V (ed) (2019) JWR Whitehand and the historico-geographical approach to urban morphology. Springer, Cham
Oliveira V (2020) The ground plan as a unifying concept. J Urban Res Dev 1:45–59
Oliveira V, Monteiro C (2014) The morphological basis of practice: learning from doing. Urban Morphol 18:72–74
Oliveira V, Monteiro C (2016a) Casa Lindo Vale. Urban Forms, Porto

Oliveira V, Monteiro C (2016b) Urban morphology, urban planning and architecture. U+D Urbanform Des 5:66–77
Oliveira V, Yaygin M (2020) The concept of the morphological region: developments and prospects. Urban Morphol 24:35–52
Oliveira V, Monteiro C, Partanen J (2015) A comparative study of urban form. Urban Morphol 19:73–92
Oliveira V, Medeiros V, Corgo J (2020) The urban form of Portuguese cities. Urban Morphol 24:145–166
Schoenauer N (1981) 6000 years of housing. W W Norton and Company, New York
Strappa G (2018) Reading the built environment as a design method. In: Oliveira V (ed) Teaching urban morphology. Springer, Cham, pp 159–184
Ünlü T (2013) Thinking about urban fringe belts: a Mediterranean perspective. Urban Morphol 17:5–20
Whitehand JWR (1967) Fringe belts: a neglected aspect of urban geography. Trans Inst Br Geogr 41:223–233
Whitehand JWR (1972) Building cycles and the spatial pattern of urban growth. Trans Inst Br Geogr 56:39–55
Whitehand JWR (2009) The structure of urban landscapes: strengthening research and practice. Urban Morphol 13:5–27
Whitehand JWR (2019) Green space in urban morphology: a historico-geographical approach. Urban Morphol 23:5–17
Whitehand JWR, Gu K, Conzen MP, Whitehand SM (2014) The typological process and the morphological period: a cross-cultural assessment. Environ Plan 41:512–533

Chapter 11
Atrium-House: An Exercise in Self-Analysis

Frederico de Holanda

Abstract In this study of an atrium-house the designer, the user and the critic are the same—therefore the title. The house was designed in 1999 and the family has since moved in. The analysis is carried out in two levels: architectural configuration, and deployment of subjects and activities in space-time. The first level is subdivided into volumetric and spatial aspects, encompassing: (a) references to history spanning from Pompeii to pre-Columbian America; (b) volumetric decomposition and form; (c) hierarchy of spatial elements and its relation to the intelligibility of the layout, among which the atrium and a long north–south axis stand to the fore; and (d) relations of physical permeability and perceptual visibility, both among internal spaces and between these and the outside public space. The types of architectural performance—functional, social, imagistic, etc.—are analysed and their relative importance in determining the architectural *partis* is indicated. In the second analytical level—space use—a critical appraisal is presented, made possible by the architectural experience of the place along its 21 years of use. It is discussed whether the 'panopticum effects' of the atrium may be differently interpreted: authoritarian surveillance of some subjects over others, or easy access and mutual visibility among people, enhancing possibilities of social contact.

Keywords Atrium-house · Architectural configuration · Architectural history · Modern architecture · Space use

11.1 Introduction

Brasília is well-known for its 'superblocks'—the residential areas of the Pilot Plan (henceforth 'Plan') designed by Lucio Costa in 1957, with apartment buildings six-stories high over *pilotis* (Fig. 11.1). It is less known for the palimpsest that has constituted its urban tissues over history (Holanda 1998): besides the modernistic fabric, the historic core of pre-existing vernacular towns (*Planaltina*, 1850; *Brazlândia*, 1930),

---

F. de Holanda (✉)
Universidade de Brasilia, Brasilia, Brazil
e-mail: fredholanda44@gmail.com

V. Oliveira (ed.), *Morphological Research in Planning, Urban Design and Architecture*, The Urban Book Series, https://doi.org/10.1007/978-3-030-66460-2_11

Fig. 11.1 A typical Brasília 'superblock' (*Source* Photograph by Frederico de Holanda)

slums (which have been bulldozed by the local government), working camps (built by contractors, of which a fascinating example survives: *Vila Planalto* [Holanda 2007]), 'post-modern' satellite towns (some of them with buildings up to 40 stories high) and gated communities (for all income levels, a very common pattern of urban expansion). The originally designed 'bit' of the city—the Plan—houses only *circa* 8% of the city's population; 92% are distributed in discontinuous urban patches, in all directions from the core (CODEPLAN 2015).

From its inception, urban policies in Brasília have: (i) prevented the occupation of the surrounding areas of the Plan so that the original 'physiognomy' of the city would remain highly legible in the landscape (Costa 1995); (ii) defined the satellite nuclei themselves as discontinues urban patches, very much in the garden-cities' spirit of urbanised areas separated by rural occupation. Over time, there has been some infilling of such interstices, but Brasília remains a very dispersed city—actually, from our surveys, the second most dispersed city in the world, only after Mumbai, India (Holanda et al. 2015). Such territorial policies contribute to high real estate prices in the core, forcing lower and even middle classes to flee the Plan in search for more affordable housing.

Before proceeding, we must consider some theoretical elements. I start from Bill Hillier's idea that architectural design is a 'conjecture-test' process (Hillier 1996) and I illustrate this with a 'virtuous circle' (Fig. 11.2): architecture must be understood in a two-fold fashion, through its 'determinations' (upper half of the figure) and through its 'effects' (lower half of the figure)—in other words, as a 'dependent' variable and as an 'independent' variable, simultaneously (for a thorough discussion, see Holanda 2018). Concerning determinations, there are the three-fold categories

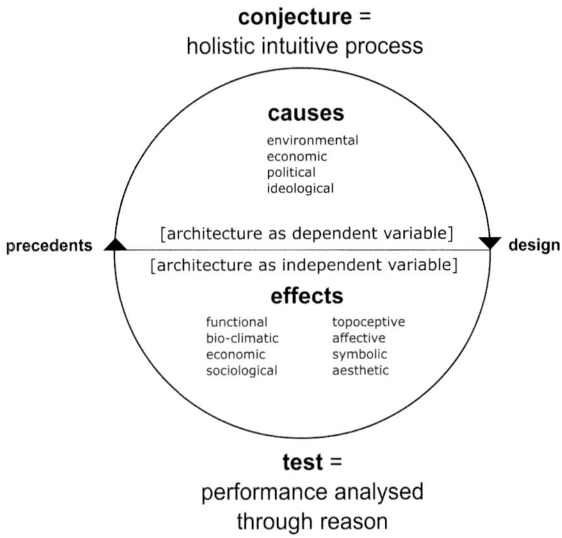

Fig. 11.2 The 'virtuous circle' of the process of design—inspired in Hillier (1996), Giddens (1984) and Popper (1976)

of the Marxian tradition—economic, political and ideological—to which I add a fourth one: environmental; concerning effects, albeit not ignoring impacts on nature, the program at our research group at the School of Architecture and Urbanism, University of Brasília, concentrates on the effects directly 'on people', in various 'aspects' or 'dimensions'—Fig. 11.2 illustrates the eight categories through which we classify such effects. Narrowing further my personal contribution to the research group, I focus on the 'sociological effects' of architecture on people, although, in the comments to follow on the house I shall refer briefly to other dimensions of performance of its architectural volumes and spaces.

I thus propose a subdiscipline within the field of the discipline of Architecture, which I call Sociological Architecture. The research endeavour strives to answer the following questions.

Does the configuration of form-space (solids, voids and their relations) imply desirable ways of individuals and groups (social classes, genders, generations) deploying themselves in places and moving through them, and accordingly desirable conditions for encounters and avoidances and for the visibility of the 'other'? Do the type, quantity and relative location of activities imply desirable patterns of utilisation of places, in space and time? (Holanda 2010).

The formulation is inspired by Space Syntax Theory (henceforth SST: Hillier and Hanson 1984; Hillier 1996; Hanson 1998), but I include the 'semantics' of architecture (labels superimposed on buildings and open spaces), besides its 'syntax' (configuration of barriers and permeability, as well as opaqueness and transparency). SST has as its fundamental axiom that 'space is a function of forms of social solidarity' (Hillier and Hanson 1984), and it may be—and has been—applied to all scales of architectural space. This chapter exemplifies this, encompassing in similar methodological procedures the analysis of a metropolis of almost 3,000,000 people and a single family house. The 'method', as constituted by 'analytical categories and their relations', will accordingly depict initially the attributes of the city (albeit a very short analysis—for a thorough study see, for example, Holanda 1998, 2007, 2011) and afterwards it will concentrate on the scale of the domestic realm. In both cases the focus will be on how architectural configuration impinges upon modes of social appropriation of places. And, finally, the techniques will indicate the 'practical procedures' through which the analysis is carried out.

Coming back to the scale of the city at large, besides being very dispersed, Brasília's street and road system is very fragmented and the majority of its segments (52%) are cut off from the surroundings, their access being privately controlled (Coelho and Medeiros 2019). Dispersion and fragmentation impose a heavy burden on the daily life of the inhabitants, in terms of distances and time-length of journeys to work and services, in particular to the poorer social layers. Zechin (2014) shows that for the poorer layers' average distance from home to the urban centre is 25.8 km, whereas for the richer ones is 4.9 km. In the case of Fortaleza, the state capital that comes in second in this score, the distances are 7.4 km and 4.4 km, respectively.

The house I examine is in one of the gated communities' layouts: a middle-class environment that belongs to a borough almost entirely constituted by such schemes—*Grande Colorado* (Fig. 11.3)—22 km away from the metropolitan core. In the borough there is a central avenue along which there are two types of occupation: (i) buildings 2 to 4 stories high, in plots without setbacks and continuous façades, with shopping in the ground level, and offices or flats in the upper one (terribly ill-configured, with parking spaces in between street sides, and poorly designed sidewalks, Fig. 11.4), and (ii) long tracts of fences of the gated arrangements (Fig. 11.5),

Fig. 11.3 The Plan and the *Grande Colorado* borough (*Source* Google Earth)

Fig. 11.4 Commercial strip of the *Grande Colorado* borough (*Source* Photograph by Frederico de Holanda)

Fig. 11.5 The long fences defining 'blind' public spaces (*Source* Photograph by Frederico de Holanda)

defining 'blind' spaces, i.e., portions of space where nothing opens, with a reduced number of entrances for vehicles and pedestrians. The option for this anti-urban landscape is an example of multiple determinations of an architectural decision: true, there came a time when we could no longer afford the prices in the Plan for the amount of built area that we needed, but add to this the architect's dream of designing a house for himself and his family with full freedom, and the lack of alternatives—bar gated communities—in urban expansion areas of the Federal District. All things considered, here we are for 21 years now (2020).

11.2 The House—Preliminaries

The technique of 'axiality' comes from SST, and indicates, through the measure of 'integration', the degree of topological accessibility of the elements of the street (and road) system of the city, representing it in terms of straight line segments—the warmer the colour of a segment, the least number of turns we have to perform, in average, from the whole system, to reach that line. Figure 11.6 shows the most complete axial map of the municipality of Brasília to date (Coelho and Medeiros 2019). Coloured in hues of green, the borough presents values of integration approximately in the centre of the interval.

The house is built on a 20 × 60 m plot, in the satellite city of *Sobradinho* (Fig. 11.7). It was designed for a nuclear family of four members: husband (myself), wife, a 48 years old daughter and a 47 years old son (2020), but daughter and son now live in other cities and come to Brasilia only for visiting, some weeks each year. A young couple work as servants to the owners; an apartment was built for them, within the

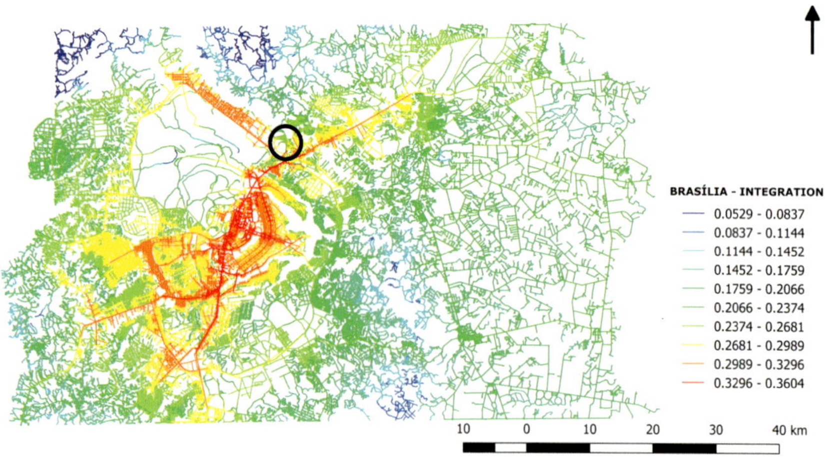

Fig. 11.6 Axial map of the municipality of Brasília (Federal District). The black circle indicates the borough

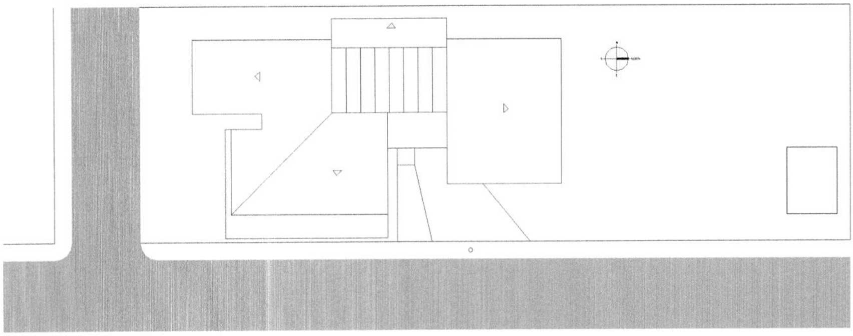

Fig. 11.7 House and plot (the rectangle to the far right in the plot is the servants' premises; in grey, the street)

plot, in 2014. The total built area is 385 m^2 if we include the atrium (50 m^2), which is open to the sky, and the servants' apartment (35 m^2), thus occupying 32% of the plot's surface. The main house is placed in the plot's southern half, thus leaving space for a generous orchard to the north, bar the servants' flat. The play of volumes pays a double tribute (Fig. 11.8): to the trapezoidal elements of Brazilian modern architecture and to Le Corbusier, in the complexity of the masses of Ronchamp Chapel, and not the purism of Ville Savoy. Finally, surface textures, colours and the play of sunlight beams in the internal space of the house recall my strong impressions of Luís Barragan's houses in Mexico City.

The house was designed after we made a second visit to pre-Columbian Mexico (1997) and it was ready on 21 June 1999, when we moved in. In that trip we knew the recently excavated atrium houses of Teotihuacan (Fig. 11.9), which in turn recalled our affections in the Alhambra, Granada, Spain (Fig. 11.10), and Pompeii, Italy, almost 20 years before. The 'palaces' of the *Avenida de los Muertos* (Avenue of the Dead) amazed me in their string of atrium after atrium—like in the Alhambra—as a spatial device that creates a larger scale, with longer perspectives and enhanced possibilities of mutual awareness and direct movement between people at the domestic realm. The bio-climatic comfort of these inner spaces has further stricken me: it was

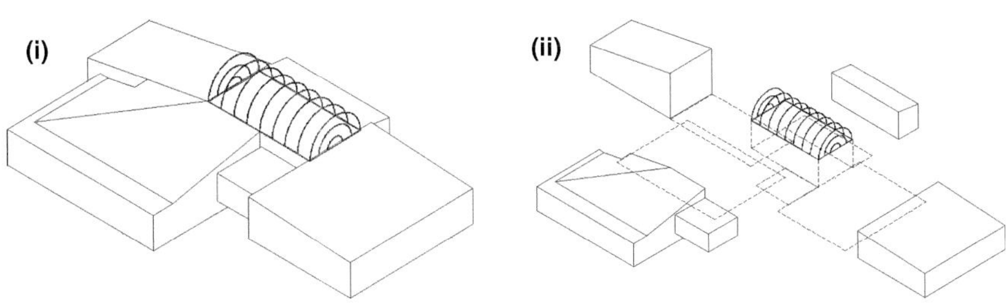

Fig. 11.8 The volumes: (i) together, and (ii) 'exploded', showing their trapezoidal form and relation to the atrium

Fig. 11.9 Archaeological remains of atrium houses of Teotihuacan: (i) model, (ii) remaining atrium (*Source* Photographs by Frederico de Holanda)

Fig. 11.10 Alhambra, Granada (Spain), Arrayanes Patio (*Source* Photograph by Frederico de Holanda)

very warm outside, in the public spaces, but because of the thermal inertia of the thick walls and the 'chimney effect' of the atriums, which pulls upwards and out of the building warmer air, inside spaces have very mild temperatures without the help of any mechanical device. In Pompeii, in turn, what fascinates me are the changes in 'isovists', both in plan and in cross-sections, along the sequence of identifiable spatial units, subtle as they are, from the public space of the street to the inner atrium, with its *impluvium* and the fountain in its centre, through a series of spatial steps (see below how I 'syntactically' re-interpreted this in our house in Brasília). The concept of 'isovist' comes from Benedikt (1979) and refers to visual polygons linking a vantage point with the edges of visible surfaces.

Open spaces inside buildings in Brazil are common, but rather in the form of courtyards in convents or other institutional buildings. 'L' or 'U' shaped houses were built from colonial times, forming gardens towards the back of the plots; however, such spaces are rarely completely enclosed and permeable to their contiguous rooms as an atrium is. In modern domestic Brazilian instances, internal open spaces are also ill-defined. Our house is closer to the models of Teotihuacan, Alhambra and Pompeii.

I suspected the atrium would match our expectations in various aspects, particularly bio-climatic, functional, aesthetic and sociological. I report on how we assess the house today and I comment on how it fits into present architectural panorama. Studies by Amorim (1997), Hanson (1998) and França (2001) are considered.

11.3 'Public' Space and the Domestic Realm

As the house is in a gated community, 'public' space should be qualified: it concerns not the 'urban public' but the 'neighbours' of a community of circa 800 people (216 houses). And yet, there are lessons that can be retrieved from the way the house relates to the exterior space, in unpredicted ways, which could apply in 'normal' urban settings (for an idea of the play of volumes, see Figs. 11.11, 11.12 and 11.13).

Consider the *promenade* from the point on the sidewalk in front of the main door to the centre of the atrium (Fig. 11.14). This sequence of spaces reinterprets that of the Pompeian house by creating isovists which change in size and form along the route, as well as it includes some attributes which are not found in the Roman city. Firstly, in the sidewalk: in the volumetric (de)composition of this house, some of the volumes present setbacks (the yellow volume, Figs. 11.12 and 11.8), some do not (the blue volume, in the same figure); the first isovist allows for generous perception of the public space towards left and right (the red wideband to the bottom of Fig. 11.14). Secondly, in the ramp: now the isovist is limited by the blue volume to the left and the yellow one to the right, further ahead; the ramp does not exist in Pompeii, but it is a syntactical separation that parallels the stairs that sometimes we find in the transition from the sidewalk to the porch in the latter. Thirdly, in the porch: this is still an open, public space, but covered by a concrete slab; let us say that it is an isovist defined in three dimensions, not two—it is indeed perceived as

Fig. 11.11 North-east view (*Source* Photograph by Frederico de Holanda)

Fig. 11.12 East view (*Source* Photograph by Frederico de Holanda)

Fig. 11.13 South-east view (*Source* Photograph by Frederico de Holanda)

a distinguished spatial step. Fourthly, in the vestibule: this is the first spatial unit in the sequence 'inside' the house. Fifthly, in the corridor, the isovist that opens wide towards north and south. And finally, in the atrium: although it is the spatial core of the house, the isovist is smaller in area than the previous, for we miss the generous transparency towards north and south provided by the corridor.

This sequence of spaces, subtly differentiated as it may be, has led to behaviours that have surprised me: (i) strangers and not close acquaintances stay on the sidewalk, they do not step on the ramp, which is paved with ceramic tiles, in contrast with the sidewalk, which is paved with stones, and from the sidewalk they ring the bell, reachable in the blue wall to the left (Fig. 11.12); (ii) closer acquaintances ring the bell and advance until the middle of the ramp or the porch; (iii) close friends and relatives (sometimes) ring the bell, climb the ramp, reach the porch, open the door (it is never locked) and enter. I argue that the three classes of subjects 'read', in space, the 'tips' to behave the way they do: as Bill Hillier has always insisted, 'spatial differentiation' carries with it, as 'intrinsic attributes, social differentiation'—this may happen in unexpected ways.

11.4 Atrium and Axis

As to the internal spaces of the house, another analytical procedure, also coming from SST, has been applied, this time to support the study of relations among spaces in two

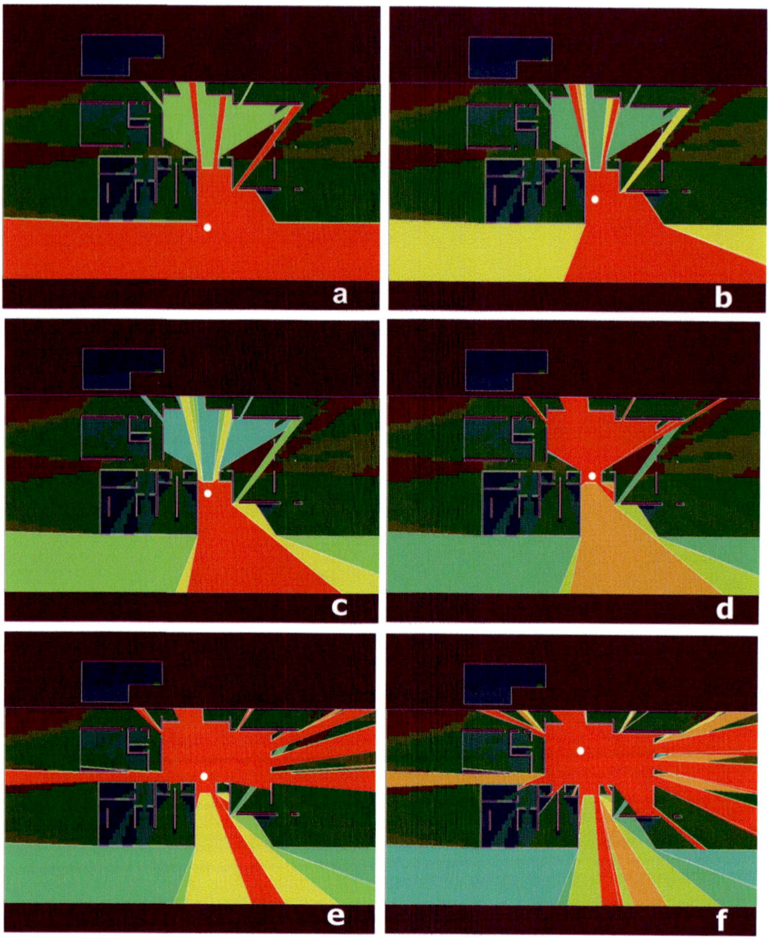

Fig. 11.14 Isovists from the sidewalk to the inner atrium. Isovists' form and area are in red; white dots indicate the point from which each isovist is drawn: (a) from the sidewalk; (b) from the ramp; (c) from the portico; (d) from the vestibule; (e) from the adjoining corridor; and (f) from the atrium (Software: Depthmap)

dimensions, not in one dimension as in the axial map: the technique of 'convexity' (Hanson 1998). The internal space of the house is obviously continuous, but we may identify, say, elementary units that we intuitively perceive as 'heres' and 'theres': these are the 'convex spaces' that represent the structure of permeability inside the house (Fig. 11.16).

Two spaces are the main elements of the *partis*: the atrium and the long north–south visual axis adjacent to it—henceforth NS-axis (Figs. 11.15, 11.16, 11.17, 11.18, and 11.19). The atrium opens wide to adjoining spaces by means of sliding glass

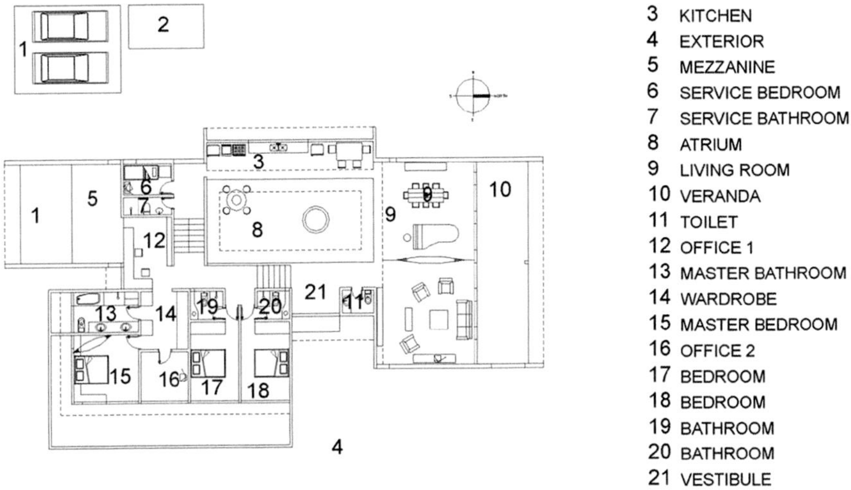

Fig. 11.15 Plan: Garage (1) and laundry (2) are in part underneath mezzanine (5), service bedroom (6) and service bathroom (7)

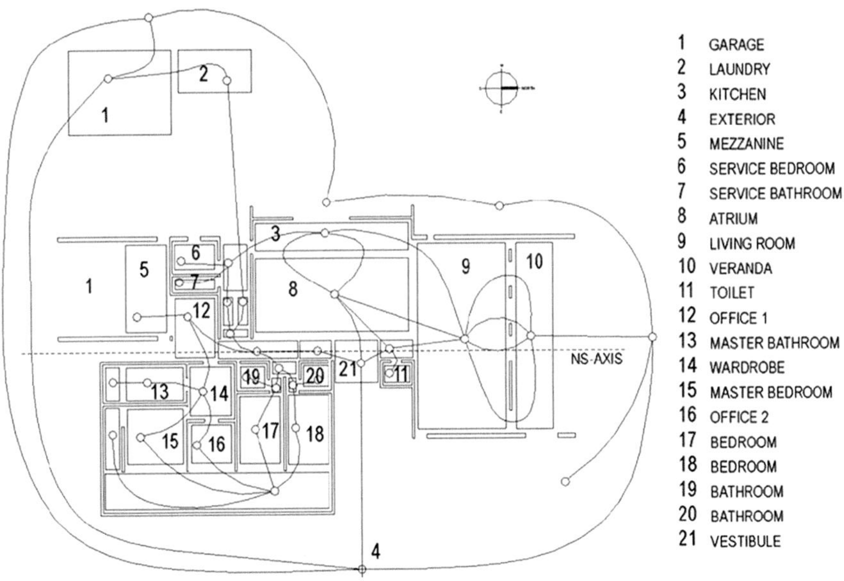

Fig. 11.16 Plan, showing 'convex decomposition' of internal space and permeability graph. Garage (1) and laundry (2) are in part underneath mezzanine (5), service bedroom (6) and service bathroom (7). Dots represent convex spaces and lines permeability among them

Fig. 11.17 (a) NS-axis viewed from office 1; (b) NS-axis viewed from living room (*Source* Photograph by Frederico de Holanda)

doors, except to the southern side, closed by a panel of decorative tiles—*azulejos*, a long Portuguese-Brazilian tradition, which has been rescued by Brazilian modern architecture. The atrium is covered by a barrel vault metal pergola, allowing for the 'passage' of sun, moon, rain, but controlling the passage of people—humming-birds often enter the space to enjoy the flowers (Fig. 11.20). The atrium sliding doors are opened ajar day and night, the year round, except in circumstances of heavy/windy rain, or exceptionally cold days in the winter, when temperatures may fall to *circa* 8 °C or less at night—a rare event in a weather which is usually mild in the Central Plateau in Brazil. Out of sheer curiosity: when I visited the *Torre de Belém* (Belém Tower), a fortress by the Tejo river shore in Lisbon, I was happy to see that the atrium at the ground level, where the canons are, has always exactly the same dimensions in plan as ours: 5 × 10 m.

The NS-axis crosses visually the house from the backyard lawn (north) to the exterior space in the street (south); it goes through the veranda, living/dinning room, vestibule, some circulation spaces and office 1. The living room (72 m^2) includes a sitting area, a professional piano—the daughter is a professional classical pianist and I am an amateur player—and a large table used in daily meals as well as on special occasions. Ideally, all spaces should turn around the atrium and axis. It did not turn

Fig. 11.18 NS-axis, atrium and kitchen (far right); reinforced concrete allows elimination of traditional peripheric columns, thus stressing spatial continuity (*Source* Photograph by Frederico de Holanda)

Fig. 11.19 Kitchen (far left), living room and doors leading to the veranda (far right) (*Source* Photograph by Frederico de Holanda)

Fig. 11.20 Atrium: pergola and decorative tiles panel by Brazilian designer Petrônio Cunha (*Source* Photograph by Frederico de Holanda)

out completely so because of economic, environmental and functional limitations—plot dimensions, room surface and number, required functional positioning, the need to adapt the house to the plot slope (*circa* 5%) without severe earth cuts, infills, etc. The idea was to maximise accessibility/visibility by means of 'synchronising' space (Hillier and Hanson 1984) in both two (atrium) and one (axis) dimensions: the atrium does it to the spaces around it, the axis does it to the spaces along its length. Synchronisation fosters intelligibility of the scheme, by introducing order over and above the scale of the elementary spatial units; even in the small realm of a single family house, it yields a 'global logic' (SST jargon) in the same way.

11.5 Domains, Not Sectors

We should not talk about 'sectors' in this house—social, service, private—in the way we traditionally do in domestic space in Brazil (Amorim 1997, Fig. 11.21). 'Justified' graphs, i.e., spaces represented according to its topological distance from the exterior, reveal how insulation between activities subsumed under such labels does not exist here: bedrooms ('private') open independently to 'social' spaces; so does the kitchen ('service') to the atrium and living room ('social'); links between 'service' and 'private' cross the circulation between 'social' and 'social' (e.g, between living room and office 1, both belonging to the 'social' sector). Relations of accessibility and visibility among elements of different 'sectors' are usually direct, without intervening circulation spaces. In contrast with traditional evidence, strings of elements of a same 'sector' may be interrupted by elements of another 'sector'. In traditional modern *partis* this would constitute a 'functional mistake'; here, it is a strategy of organising social categories in a different way.

Rather, we should speak of two 'domains': 'non-restricted' and 'restricted' (Fig. 11.22). The non-restricted domain is constituted by spaces in which people may interact for purposes which do not involve intimacy: talking, watching TV, listening to music, preparing and eating meals, washing clothes or dishes, receiving friends, etc. These spaces are open: no doors ever shut them off from the rest of the house. This is the domain of the atrium, living/dining room, kitchen, office 1, laundry and garage.

The restricted domain is constituted by spaces for intimacy, for inhabitants or visitors: personal hygiene, talking confidentially, sleeping together, making love. All such spaces are closed: they may be shut off from the rest of the house by doors—no

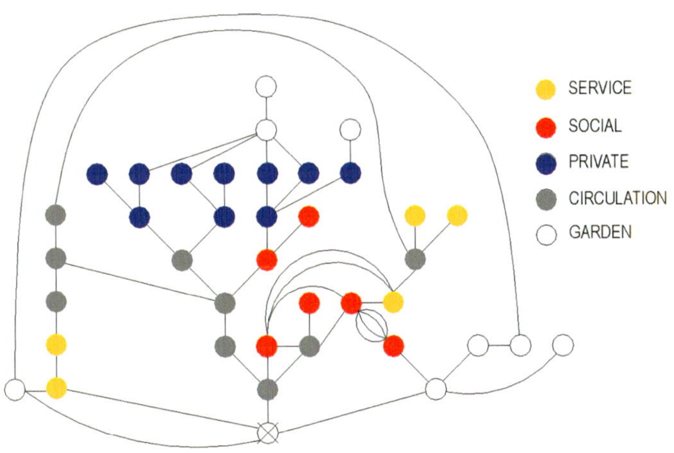

Fig. 11.21 Justified graph from exterior—spaces classified according to traditional 'sectors' (dots represent convex spaces, lines represent permeability among them, colours—see caption—represent 'sectors')

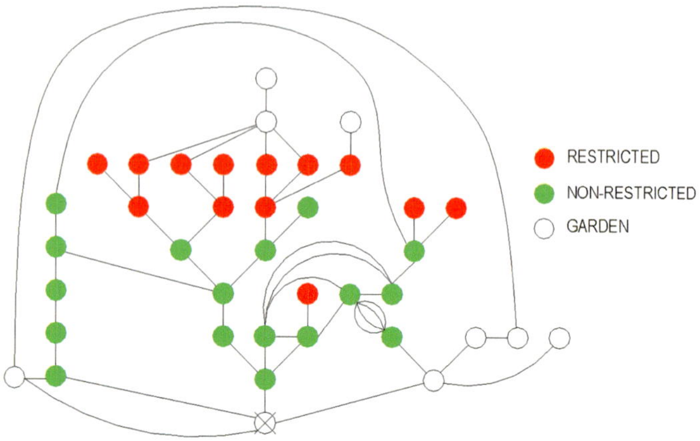

Fig. 11.22 Justified graph from exterior—spaces classified according to 'domains' (dots represent convex spaces, lines represent permeability among them, colours—see caption—represent 'domains')

exception. This is the domain of the suites, bedrooms, bathrooms/toilets (including the one adjoining the living room). This classification mixes social categories, cuts across usual dichotomies—inhabitant/visitor, owners/servants, older/younger and male/female.

Boundary/convex ratio (0.28) captures such characteristics (this ratio is the percentage of bounded spaces over the total number of convex units of the building—Hanson 1998). This house is much more open than modern architects' houses studied by Amorim (1997): they average 0.41, and only two of his eight houses are more open than this one. London's sample of architects' houses studied by Hanson (1998) presents an even greater degree of closure (average = 0.42) and only two out of eighteen are more open than this.

França's sample of middle-class 'social knowledge' houses in Brasilia is very similar to both Amorim's and Hanson's examples: average boundary/convex ratio is 0.40, and only two (out of 27) invest in higher openness than this house. Still, this is rather surprising: I would expect 'vernacular' houses to present less spatial continuity (França 2001).

11.6 Integration and Inversion

The overall measure of convex integration, i.e., the degree in which a space (a convex unit) is topologically accessible to all others in the system, also distinguishes this house from others designed by architects, but differences from vernacular are more peculiar. Average integration is 0.84, including exterior space.

Architects generally prefer deeper schemes (if there is in average a great number of intervening spaces among convex units, we say the scheme is 'deep'—i.e., low integration measure; otherwise, it is 'shallow'—i.e., high integration measure). Average integration in Amorim's sample is 0.76 and only one of his houses is more integrated than this: 1.13. The strategy replicates with Londoner architects: a sample of eighteen instances presents 0.70 as average integration, and only two examples are shallower than this house. When we turn to some houses designed by 'architectural stars', average integration varies from 0.50 (Loos' Muller House) to 0.72 (Meyer's Giovannitti House) (Hanson 1998).

The picture changes when we turn to vernacular. França's sample presents an average integration of 0.85. This poses this house closer to 'social knowledge' than to 'professional knowledge', in this variable. However, uniqueness is revealed as to the integration of specific spaces.

Consider the 'main spaces' of a house: living room, kitchen, main bedroom, exterior. The order in which they are organised from least to most integrated says a lot more than average integration. In França's sample, no house presents the exterior space as the most integrated one. In almost half the examples, integration rank order is: living room > kitchen > main bedroom > exterior; in 70.7% of cases the exterior is the 'most segregated' space. In Amorim's sample, exterior space is always among the most segregated nodes of the house. In this house, it is the most integrated one: rank order is exterior > living room > kitchen > main bedroom. Socially, this implies that, as a whole, this house actually 'inverts' the relations between inside spaces and outside ones—i.e., between domestic and public realms—as compared to both professional and social knowledge.

Amorim has also experimented with 'visual integration': relations of 'visibility' among spaces, not physical 'accessibility' as in the traditional integration measure. Visibility brings this house closer to his sample: it is still more visually integrated (2.63) than Amorim's average (2.03), but now three out of his eight cases are more integrated than this house. As in the reasoning above, more peculiar attributes appear when we consider specific spaces.

Accessibility integration, compared with visual integration obtained by the Depthmap software, illustrates this. I have run the program (i) for interior spaces only (Fig. 11.23); (ii) including exterior spaces of the immediate surroundings (Fig. 11.24), in order to capture the way relations of visibility happen between inside and outside worlds, by means of glass surfaces. Notice how atrium, living room and NS-axis belong much stronger to the integration core in visibility integration than in accessibility integration. Another specificity of this house is revealed: its transparency towards exterior space, against the opaqueness of atrium buildings which were its inspiration—Teotihuacan, Alhambra, Pompeii.

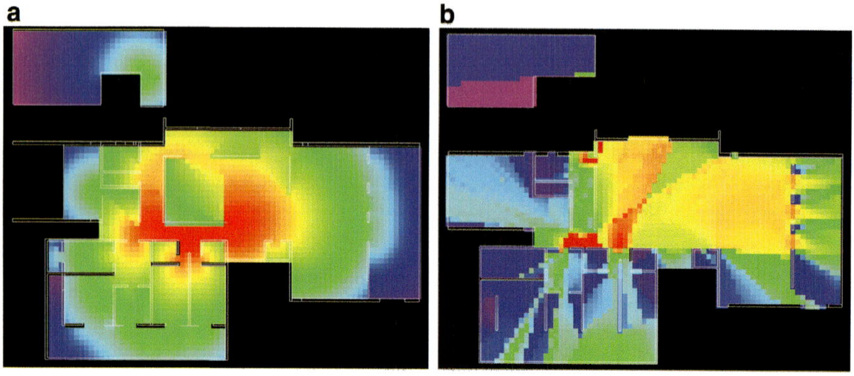

Fig. 11.23 (a) Accessibility integration compared with visual integration (b) interior only (tessellation of 0.4 × 0.4 m)

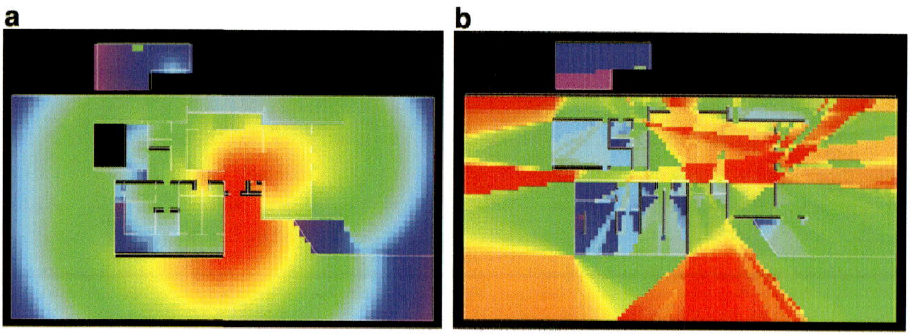

Fig. 11.24 (a) Accessibility integration compared with (b) visual integration—including exterior (tessellation of 0.7 × 0.7 m)

11.7 Co-presence and Co-awareness

The reasoning above implies that we must distinguish 'co-presence' from 'co-awareness': the first concerns the simultaneous presence of people in any convex unit of the system; the second is the perception of people in other spaces, either by sight or hearing, in places directly accessible physically or otherwise.

In daily usage the most intensely occupied space is the living room, in reading, listening to music, both preferably in a hammock, a very traditional cloth pendent device for resting, reading or sleeping in Brazil (foreground, Fig. 11.18), positioned almost exactly along the NS-axis, or watching TV in the sitting area. Daily meals are also made at the table in the living room. We are far away from Brasilia's middle-class intense use of the kitchen as the place *par excellence* for family daily interaction, which goes together with a formal living room exclusively used for receiving guests. The second space more intensely used is office 1, by both me and my wife, in work (more usually), writing letters or leisure—all involving computers.

Opposite ends of the NS-axis within the house—places of longest stay—are privileged positions concerning both visibility (more) and accessibility (less)—Figs. 11.23 and 11.24. It is also easy to listen to what is going on in the house, and to talk to people located in any bit of the non-restricted domain—even if you are not seeing him/her. Here a quasi-problem: in this domain there is no acoustic privacy: volume on the TV set or sound equipment has to be negotiated in order to allow, e.g., for simultaneous watching TV (living room) and work (office 1). Metric distance helps. Negotiation usually succeeds…

When we receive guests, the most used place depends on the number of people involved and the weather—quite dry from May to September and mild all year round. The usual picture is: small numbers = bigger table at the living room; medium numbers = atrium + bigger table at the living room; large numbers = atrium + all living room area (table + sitting area).

Receiving guests often involves meals, prepared by inhabitants or friends who also enjoy doing so. During preparation, presence splits between kitchen, atrium and living room; contiguity/visibility/openness allows for easy talk among everybody. In the meantime, visiting children enjoy intersecting spatial rings in hide-and-seek games (bedrooms are kept open) or play with computers in office 1.

The syntax of the house subverts common classifications among 'social', 'private' and 'service' categories, shuffling subjects and practices carried out in the sectors; the transparency of the kitchen towards atrium, living room and even part of the bedrooms has posed an unanticipated issue, one that qualifies space depending on the subjects involved. Along these 21 years of inhabiting the house we have had people working for us who varied from (previously) strangers to close relatives. In the case of strangers—simply a working relationship of employer/employee—transparency implied an uneasy surveillance over their job; in the case of family ties, on the contrary, it implies a desirable proximity, furthermore in cooperative tasks often done together with the owners. As Hanson observes, we must be cautious against over-simplistic generalisations and [point] to the difference which exists between describing the objective properties of a house and the subjective interpretation of what those properties might mean either experientially or socially (Hanson 1998, p. 273).

Transparency does not always mean 'inclusion', separation does not always mean 'exclusion': the 'semantics' of subjects bears upon social relationships as much as the 'syntax' of their bodies.

Socially, visual integration with the exterior stresses the awareness of the public space from inside rooms—and vice versa—and the most integrated areas inside the house enhances co-awareness among people moving about: I am very pleased to work at office 1 as well as reading in the hammock in the living room in 'red areas' of the scheme…

11.8 Conclusion

Comparative analysis reveals both similarities and differences between this house and current architectural panorama, professional or otherwise: (i) average integration brings this house closer to 'social knowledge' (higher integration) than 'professional knowledge' (deeper schemes); (ii) integration rank order (e.g., high integration of exterior space) differs from both 'social' and 'professional' knowledge, in middle-class houses, here and abroad; (iii) the shallowness of the 'private sector' is also unique, at least concerning Brazilian domestic space, vernacular or professional (tree-like schemes abound); (iv) spatial openness is not unique to this house but places it in the top layer of most open schemes; and (v) inexistence of a formal living room distinguishes it from common middle-class houses in Brazil.

Aesthetically, this house may be said to be 'modern' without great controversy: pure volumes, primary colours, generous glass panels made possible by use of reinforced concrete, etc.; syntactically, relationships with modernity are to be qualified according to specific attributes. Its strongest peculiarity is enhanced relations of visibility/accessibility among diverse categories of spaces—inner/inner spaces or inner/outer spaces.

The house responds well to the expectations we had: openness in the non-restricted domain and the possibility of reclusion in the restricted domain, allows for desired balance between co-presence/co-awareness and isolation. If this is perhaps the house's greatest quality, it may be its major flaw: the building does not seem robust enough to support other ways of behaving.

For one, the house is not 'children-proof': openness in the non-restricted domain, with corresponding visual and acoustic attributes, makes it untenable for simultaneous work and the playing around of kids; the open kitchen may be dangerous, for it is not always possible to keep surveillance over kids and stop them from having access to dangerous equipment.

Isolating the kitchen from sight/access by walls/doors would be rather easy, construction-wise; creating a formal living room (and a corresponding intimate one), or implementing a tree-like scheme in the 'private' sector would be impossible without major structural changes.

Some expectations towards architecture are more universal. Others are culturally determined or extremely specific. Family expectations concerning domestic space may be strongly idiosyncratic. This house responds well to desired social interfaces. For us, it has a high 'use value'. I am not that optimistic about its 'exchange value' in present market conditions and middle-class values…

References

Amorim L (1997) The sectors paradigm. Unpublished PhD thesis, University College London, UK

Benedikt M (1979) To take hold of space: isovists and isovist fields. Environ Plan B: Plan Des 6:47–65

CODEPLAN, Companhia de Desenvolvimento do Planalto Central (2015) Pesquisa Distrital por Amostra de Domicílios—Distrito Federal—PDAD/DF 2015. CODEPLAN, Brasília. http://www.codeplan.df.gov.br/wp-content/uploads/2018/02/Resumo-PDAD-Distrito-Federal.pdf. Accessed 7 August 2018

Coelho JM, Medeiros V (2019) Centralities in Brasília: and inequality relation. International Space Syntax Symposium, Beijing

Costa L (1995) Lucio Costa: registro de uma vivência. Empresa das Artes, São Paulo

França FC (2001) Meu quarto, meu mundo: configuração espacial e modo de vida em casas de Brasília. Unpublished MSc Thesis, Universidade de Brasília, Brazil. www.fredericodeholanda.com.br. Accessed 26 June 2020

Giddens A (1984) The constitution of society: outline of a theory of structuration. Polity Press, Cambridge

Hanson J (1998) Decoding homes and houses. Cambridge University Press, Cambridge

Hillier B (1996) Space is the machine. Cambridge University Press, Cambridge

Hillier B, Hanson J (1984) The social logic of space. Cambridge University Press, Cambridge

Holanda F (1998) The morphology of Brasília: an evaluation. Fifth International DOCOMOMO Conference, Stockholm, 16–18 September 1998

Holanda F (2007) Be aware of local properties. International Space Syntax Symposium, 6, 2007, Istambul. In: Proceedings. Istanbul: ITU Faculty of Architecture, pp. 082-01–082-13

Holanda F (2010) Sociological architecture: a particular way of looking at places. J Sp Syntax 1(2):337–355

Holanda F (2011) Exceptional space. Brasília: FRBH. E-book. Available at: https://www.academia.edu/8586469/EXCEPTIONAL_SPACE_1997_2011_complete_. Accessed 11 January 2019

Holanda F (2018) Inserting urbanity in a modern environment. In: Oliveira, V (ed) Teaching urban morphology, Springer, Cham, pp 185–203

Holanda F, Medeiros, V, Ribeiro, R, Moura A (2015) Brasília: fragmented metropolis. International Space Syntax Symposium, London. http://www.sss10.bartlett.ucl.ac.uk/wp-content/uploads/2015/07/SSS10_Proceedings_051.pdf. Accessed 16 October 2016

Popper K (1976 [1963]) Conjectures and refutations—the growth of scientific knowledge. Routledge and Kegan Paul, London

Zechin PAV (2014) Sobre a dimensão espacial da desigualdade social urbana: um estudo sobre cinco cidades brasileiras. Unpublished PhD thesis, Universidade de Brasília, Brazil

Chapter 12
Morphological Research in Practice

Vítor Oliveira

Abstract This chapter collects some of the main issues and arguments of the book. The first part presents urban morphology as a framework to rigorously describe urban form at different scales of analysis, identifying the main agents and explaining the fundamental processes that change, or conserve, that physical form over time. The second and third parts explore the relation between the science of urban form and professional practice. Recognizing the gap between knowledge and action, the focus is on the integration of morphological approaches into action on the urban landscape and on the crucial role of the researcher-practitioner. Although the book argues for an evidence-based practice, it does not aim to produce a uniform product. On the contrary, it is expected that the diversity and richness of contexts that are acknowledged by urban morphology might, in fact, open the range of planning, urban design and architectural solutions, many times narrowed by ephemeral global trends and fashion. Finally, the specificity of planning and urban design, on the one hand, and architecture, on the other hand, is explored. While the relation between plan preparation and implementation, the multiple dimensions of planning processes and evaluation in planning are key topics of the former, the frame of references for design, the effective integration of a new object into the extant landscape, and the development of minimal approaches are fundamental themes of the latter.

Keyword Urban landscape · Urban morphology · Researcher-practitioner · Urban planning · Urban design · Architecture

12.1 The Science of Urban Form

Urban morphology is a field of knowledge with more than one century of disciplinary history (see Whitehand 1981 for the oldest tradition in the study of urban form). Throughout this period, the field has conceived, tested and refined—through application in different geographical contexts—a number of theories, concepts and

V. Oliveira (✉)
Universidade do Porto, Porto, Portugal
e-mail: vitorm@fe.up.pt

V. Oliveira (ed.), *Morphological Research in Planning, Urban Design and Architecture*, The Urban Book Series, https://doi.org/10.1007/978-3-030-66460-2_12

methods to describe and explain the physical form of cities, while understanding the main agents and processes that shape it. Why is this body of knowledge relevant to planning, urban design and architecture?

Contrarily to urban morphology, these three related fields are mainly oriented to prescription and action on the urban landscape. Their normative and proactive nature can be translated into transformation, but also into conservation (see, for instance, the Bologna plan, coordinated by Cervellati in the late 1960s), of urban forms. Paradoxically, both guidance offered by plans and direct action framed by projects are not supported by scientific knowledge on the object at hands. Addressing that weakness, urban morphology can offer professional practice a comprehensive understanding of the urban landscape by describing and explaining the territory as a whole and in its different parts, considering its evolution through time, and identifying the needs of conservation and change of the different parts—balancing social, economic and environmental aspirations with the maintenance of character and identity.

The tension between conservation and change is a fundamental theme of debate. One might provocatively frame such debate, as Karl Kropf did in Chapter 5, by stating that there is nothing in the existing state of cities that determines how they should be built in the future. We might even question the maintenance of some elements of urban form. Indeed, why should conservation be a concern? As MRG Conzen argued (and Jeremy Whitehand reminded us in Chapter 2) the urban landscape embodies not only the efforts and aspirations of their present occupiers but also those of their predecessors; and this enables people to take root in an area, acquiring a sense of the historical dimension of human experience. For Conzen, historical urban landscapes were assets to society both intellectually and as emotional experiences. All these dimensions are crucial for understanding the tension between conservation and change. All territories, and each part of each territory (old or new, central or peripheral), is subject to preservation and transformation. But, what to conserve and what to change? How to select? Can the reasoning behind selection be made explicit, as it means favouring certain histories and meanings? And, how to effectively carry out specific processes of conservation and change? While most chapters implicitly deal these questions, Chapters 3 and 10, by Peter Larkham and Nick Morton and by Vítor Oliveira and Cláudia Monteiro respectively, explicitly address them. The former frames the conceptual debate; the latter pragmatically identifies, in the project of a house, which elements have been preserved and which have been transformed, and the reasons for selection.

The aim of integrating morphological research into practice is not to promote one architectural style or one planning perspective, but to make explicit how a morphological view of cities can lead to the design of policies, plans and projects, and considering how they can be better suited to the urban landscapes to which they relate, and to the needs of the citizens living and working in these landscapes. To illustrate this statement, the book gathers very different plans and, particularly, very different buildings, from the Terny cemetery designed by Giuseppe Strappa to the Brasilia house designed by Federico de Holanda. Morphological research itself is not a homogeneous body. It is made of several approaches which, depending on the problem at hands, can be used in isolation or combination. Chapters 2 and 4, by Jeremy Whitehand and Ivor Samuels, respectively, illustrate this statement. Whitehand explores the practical applicability of the Conzenian approach, with a focus on the concept of morphological region (Oliveira 2019; Oliveira and Yaygin 2020; Whitehand 2009). Two aspects of his chapter must be underlined. The first is the remarkable diversity of institutional contexts framing Whitehand' s application of Conzenian research in practice, from UNESCO to a British parish. The second is the rigour of the method of morphological regionalization and the exceptional clearness of the passage from morphological description to planning prescription. On the other hand, illustrating combination, Samuels proposes an eclectic view, bringing together concepts and methods of the historico-geographical, process-typological and configurational approaches with New Urbanism ideas.

Bridging the research-practice gap in urban landscape, as in many other fields, is not an easy or straightforward task. There is a need to conceive and launch relevant links between these two separate worlds and, particularly, to follow-up after the realization of each bridging event to build an effective process extended in time. In this exploratory context, the researcher-practitioner has a fundamental importance. Almost all authors in this book fall into this figure. Almost all spend a significant part of their time teaching and researching; but they also develop a consistent practice on the urban landscape. When designing a plan or a project, they apply their frameworks of research, test the robustness of morphological theories, concepts and methods, and explore the crucial moment of translating analysis into design. When coming back to academia, they bring many concerns of 'real-world', often simplifying their tools, improving communication skills and strengthening the connection between urban form and other dimensions of urban life.

If professional practice does not have the analytical concepts and methods to address the urban landscape, and if urban morphology has demonstrated the potential of application of its tools, why is there a gap between science and practice? And how can it be overcome? Firstly, as stated in Chapter 1, the gap is not exclusive to the urban landscape fields and has been widely debated across the span of social, physical and life sciences. Secondly, the gap is not new and its roots, in some fields, can be traced back to the nineteenth century. Thirdly, as argued in Chapter 1, the activity of researchers and practitioners have different frameworks, objects, purposes, processes and results. As such, there is a need of academia to know the different aspects of professional practice. The next sections address specifically these questions.

12.2 Morphological Research in Planning and Urban Design Practice

A crucial aspect for the integration of morphological theories, concepts and methods in planning and urban design practice is effective acknowledgement of what practice is. The first chapters of the book (2–5) offer invaluable contributions to this understanding. Three interrelated themes should be highlighted—the differences between plan (or other prescriptive document) preparation and implementation, the complexity of planning processes and the evaluation of planning.

Often, plan preparation and implementation are two distinct things. The former is carried out by a planning team that can include the participation of external consultants. These consultants can be researcher-practitioners, as illustrated in the book. The 'product' delivered by this team, the plan, should incorporate the main elements of its planning system, favour the realization of the main needs and aspirations of the territory and population it addresses, be coherent both in its different parts and in the way it relates to other plans for the same territory and, finally, acknowledge participation of different stakeholders. Despite the complexity and the long time periods involved in plan making, the boundaries of such an exercise are relatively defined. Plan implementation has a different nature. It is often developed by a different team, many times without the participation of those consultants involved in plan making. While framed by the plan, the activities addressed throughout its implementation have a wider nature.

Indeed, planning processes bring to stage a number of aspects that are only implicit in plan making. In general, it is in this process that the tension between technical knowledge and political power is more expressive. What is the value and relevance of the plan? Throughout the process, the plan has to be useful for decision-making and consensus-building. Larkham and Morton report on how their framework for conservation areas has offered a robust, evidence-based basis that has been continuously used in decision-making, even in times of profound changes in the planning context. The availability of human and financial resources to implement plan proposals is another critical issue. In Chapter 4, Samuels addresses both the availability of adequately trained staff and economic feasibility, particularly in the Asnières plan, where the involvement of students is also stressed. Finally, the planning process is also the stage where the plan is confronted with applications, making explicit the confrontation between different agents and strategies. And as Matteo Ieva and, particularly, Giuseppe Strappa (due to the long time and the different stages of construction of the Treni cemetery) remind us, architecture is also a process.

How successful the plan was? To what extent have its proposals been implemented? Why were some proposals implemented and others not? What was the impact of the implemented proposals? What proposals have produced satisfactory results, and what went wrong? What aspects can be repeated in a future revision of the plan elsewhere? While the relevance of these questions is obvious, they are rarely asked. In Chapter 4, Samuels brings the evaluation of planning to the centre of debate. In particular, he describes how the evaluation of the Asnières plan (further detailed by Samuels and Pattacini 1997) has allowed a learning process that has then framed the development of the design code for Oxford.

12.3 Morphological Research in Architectural Practice

The frame of references for each design work is always a combination of internal and external elements. While the former is part of 'that' specific urban landscape, the latter integrates the designer knowledge and values. In architecture, more than in planning and urban design, the frame of references is usually dominated by external references. That approach is fed by architectural schools, architectural prizes and magazines, and the media. The wide acknowledgement of 'creativity' as a fundamental characteristic of architectural practice is in line with the broad celebrity status that architects have acquired in the twentieth century, the most creative of these becoming icons in mass culture (Larkham and Conzen 2014). This book proposes an alternative approach, aiming at achieving a more balanced frame of references. The specific nature of these references should be established in each particular case by morphological analysis. The internal references of an urban landscape suggested by research can have a high level of detail, including particular built forms and materials—as in the case of the Huangzhuang houses designed by Wowo Ding and her colleagues—or they could be more structural, and less visible—as in the Porto house designed by Oliveira and Monteiro.

In addition to this illustration of the range of internal references, the Porto house, presented in Chapter 10, raises another important issue for discussion. Can the architectural repertoire of the Modernist Movement fit into nineteenth-century tissues—the same tissues that have motivated its radical attitude against history? For Oliveira and Monteiro, the answer is yes. Based on morphological analysis, they develop a conservative view of that place in terms of structural elements, including the relation between building height and street width, the plot width (making evident the high urban potential of these kind of plots), the coincidence between plot and building frontages, the overall interior organization of rooms in a narrow building, and the position of the staircase within this type of building. On the contrary, they promote change of architectural style, façade design and construction materials. The result is a modernist row-house for this particular tissue. That statement is reinforced by the presentation of two other projects for Porto fitting into two other tissues, where the introduction of this architectural style would be intrusive.

How to effectively integrate a new building into an extant landscape? There is not a definite answer to this question. Yet, the third part of the book offers interesting perspectives for debate. This is particularly evident for Chapter 6. The Huangzhuang village is an exceptional settlement built on a delicate balance between water and land. The building fabric of the village is made of less than 200 buildings (including other complementary buildings) for housing and agricultural activities. Morphological and typological analysis carried out by Ding and her colleagues has revealed the structural characteristics of this landscape: the central role of the canal surrounded by two rows of houses, a number of small plots each including one main building and other complementary buildings for kitchen and bathroom, a tripartite division of the main building and the central role of the *tangwu*. Ding project conserves all these structural elements as well as the building forms and construction materials (bricks and wood), and offers effective answers to contemporary needs of residents—a crucial aspect of built heritage debate. The project gives a more permanent character to complementary buildings, provides 'light' connections between buildings and offers, through space and material qualification, new possibilities of use to the main buildings.

The houses in Huangzhuang, Quebec City and Porto share a minimal approach. While the rehabilitation of the Chinese village is based on a small number of actions in each plot, the new buildings in the Canadian and Portuguese cities are erected on small plots in nineteenth-century tissues, of intense urban life. In both cases, Quebec and Porto, plot frontage and building frontage are coincident. In Porto, the new building is on a 5 × 40m plot located in the long side of the street-block. The 5m wide plot is a usual plot in nineteenth-century Porto. This type of plot corresponds to a three bays building, of varying storeys, and can have both single-family and multi-family utilizations. In Quebec City, the new building is on a 12 × 28m plot located on the corner of the street block. The new building assumes that exceptional position in the street block, incorporating in the short side of the plot a bipartite division and in the long side of the plot a four-partite division. Both cases represent the incorporation of lessons learned from the past in terms of rational use of land and efficiency of resources use—an issue that should be part or practitioners' agenda for the next decade (Oliveira and Vaz 2020).

The 'house', as a programme, always had a central role in architectural research, teaching and practice. Within it, the design of the architect's own house presents an additional interest. Chapters 10 and 11, by Oliveira and Monteiro and by Holanda respectively, offer two cases where the client and the owner of the house coincide. These are two very different cases. One is in Europe, in an urban context, in a nineteenth-century street and in a narrow and deep plot. The other is in South America, in a gated community located 20 km of iconic Brasilia, in a twentieth-century street in a medium-size plot. The design of the former is based on historico-geographical analysis, the project of the latter draws on configurational analysis. Both cases are personal reflections on the act of 'living a house', reflections of an ex-ante and on-going (six and 21 years of life in the house) nature, on the interaction between human beings and urban forms at the most basic level.

12.4 Final Reflections

The relation between research and practice has been part of the debate on the urban landscape for many years, occasionally occupying a central role, sometimes being replaced in the centre of discussion by other themes. It is expected that the research-practice relation should continue to be debated in the next years. The relation is not easy and the consistency of achieved links, might sometimes disappoint the promoters of bridging processes. Yet, these processes should continue to be developed as they have several advantages for both sides.

What is the main contribution of this book to debate? It offers evidence on the crucial position of the researcher-practitioner. Moving between two different worlds, the researcher-practitioner has the possibility to apply in real contexts of design, the morphological theories, concepts and methods he develops in academia. Each application is an opportunity to make evident the relevance of urban morphology to planners, urban designers and architects—and also to politicians and citizens—who probably did not have a morphological education or training. The usefulness of each morphological input must be evident to these professionals for it to be integrated in their routines. This is a challenge and an opportunity to change professional culture. Change can be slow but results on the urban landscape can be considerable.

To build a process where rigorous morphological analysis supports the design of plans and projects, that are evaluated (including assessment of different design alternatives) and monitored throughout implementation and construction, bearing in mind the major needs and aspirations of both citizens and cities, in times of limited resources. This should be a major challenge for the next decade.

References

Larkham PJ, Conzen MP (2014) Agents, agency, and urban form: the making of the urban landscape. In: Larkham PJ, Conzen MP (eds) Shapers of urban form: explorations in morphological agency. Routledge, New York, pp 3–23

Oliveira V (ed) (2019) JWR Whitehand and the historico-geographical approach to urban morphology, Springer, Cham

Oliveira V, Vaz R (2020) Is the concept of urban metabolism useful for planning practice? Eur Plan Stud 28

Oliveira V, Yaygin M (2020) The concept of morphological region: developments and prospects. Urban Morphol 24:35–52

Samuels I, Pattacini L (1997) From description to prescription: reflections on the use of a morphological approach in design guidance. Urban Des Int 2:81–91

Whitehand JWR (1981) Background to the urban morphogenetic tradition. In: Whitehand JWR (ed) the urban landscape: historical development and management; papers by MRG Conzen. Institute of British Geographers/Academic Press, London, pp 1–24

Whitehand JWR (2009) The structure of urban landscapes: strengthening research and practice. Urban Morphol 13:5–27

GPSR Compliance

The European Union's (EU) General Product Safety Regulation (GPSR) is a set of rules that requires consumer products to be safe and our obligations to ensure this.

If you have any concerns about our products, you can contact us on ProductSafety@springernature.com

In case Publisher is established outside the EU, the EU authorized representative is:

Springer Nature Customer Service Center GmbH
Europaplatz 3
69115 Heidelberg, Germany

Batch number: 10756376

Printed by Printforce, the Netherlands